Facebook®

8th Edition

by Carolyn Abram
with Amy Karasavas

A Wiley Brand

Facebook® For Dummies®, 8th Edition

Published by **John Wiley & Sons, Inc.**, 111 River Street, Hoboken, NJ 07030-5774, www.wiley.com

Copyright © 2021 by John Wiley & Sons, Inc., Hoboken, New Jersey

Published simultaneously in Canada

For general information on our other products and services, please contact our Customer Care Department within the U.S. at 877-762-2974, outside the U.S. at 317-572-3993, or fax 317-572-4002. For technical support, please visit https://hub.wiley.com/community/support/dummies.

Wiley publishes in a variety of print and electronic formats and by print-on-demand. Some material included with standard print versions of this book may not be included in e-books or in print-on-demand. If this book refers to media such as a CD or DVD that is not included in the version you purchased, you may download this material at http://booksupport.wiley.com. For more information about Wiley products, visit www.wiley.com.

Library of Congress Control Number: 2021934616

ISBN: 978-1-119-78210-0; ISBN (ePDF): 978-1-119-78211-7; ISBN (ePub): 978-1-119-78213-1

Manufactured in the United States of America

SKY10027349_052821

Contents at a Glance

Introduction 1

Part 1: Getting Started 5
CHAPTER 1: The Many Faces of Facebook 7
CHAPTER 2: Adding Your Face 23
CHAPTER 3: Finding Your Way Around 35

Part 2: Day-to-Day Facebook 49
CHAPTER 4: Reading News Feed and Posting 51
CHAPTER 5: Timeline: The Story of You. 83
CHAPTER 6: Understanding Privacy and Safety. 109
CHAPTER 7: Facebook on the Go 157

Part 3: Connecting with Friends 197
CHAPTER 8: Finding Facebook Friends 199
CHAPTER 9: Just between You and Me: Facebook Messenger 211
CHAPTER 10: Sharing with Facebook Groups 231

Part 4: Getting the Most from Facebook 257
CHAPTER 11: Filling Facebook with Photos and Videos 259
CHAPTER 12: Buying, Selling, and Fundraising. 285
CHAPTER 13: Scheduling Your Life with Events 307
CHAPTER 14: Creating a Page for Promotion. 319
CHAPTER 15: Using Facebook with Games, Websites, and Apps 355

Part 5: The Part of Tens 373
CHAPTER 16: Ten Ways to Make the Most of Your Facebook Content 375
CHAPTER 17: Ten Ways to Be Politically Active on Facebook 383
CHAPTER 18: Ten Frequently Asked Questions 389

Index 397

Table of Contents

INTRODUCTION 1
About This Book 1
Foolish Assumptions 2
Icons Used in This Book 3
Beyond the Book 3
Where to Go from Here 3

PART 1: GETTING STARTED 5
CHAPTER 1: **The Many Faces of Facebook** 7
So What Is Facebook, Exactly? 8
Discovering What You Can Do on Facebook 10
    Connecting with friends 10
    Discovering what's going on with your friends 11
    Establishing a timeline 11
    Communicating with Facebook friends 12
    Sharing your thoughts 12
    Sharing your pictures and videos 13
    Planning events 14
    Joining and creating groups 14
    Using Facebook around the Internet 14
    Promoting a business 15
    Fundraising for a cause 15
Keeping in Mind What You Can't Do on Facebook 16
    You can't lie 17
    You can't be 12 or younger 17
    You can't troll, spam, or harass 17
    You can't upload illegal content 18
Realizing How Facebook Is Different from Other Social Sites 18
Finding Out How You Can Use Facebook 19
    Getting information 19
    Keeping up with long-distance friends 19
    Moving to a new city 20
    Getting a job 20
    Throwing a reunion 20
    Finding a happily ever after 21
    Entertaining yourself and playing games 21
    Communicating in times of trouble 21

CHAPTER 2: **Adding Your Face** 23
    Signing Up for Facebook...23
    Checking Your Inbox...25
        Confirmation 25
        Email outreach..25
    Getting Started...26
        Step 1: Adding a profile picture................................26
        Step 2: Finding your friends...................................28
        Step 3: Getting to know your privacy settings..................29
    Introducing Your New Home Page 29
    Adding More Friends...31
    Filling out Your Profile Information 32

CHAPTER 3: **Finding Your Way Around** 35
    Checking Out the Top Bar 36
    Search 40
    Viewing Stories and News Feed...41
        Stories 41
        News Feed 41
    The Left Sidebar..42
        The top section 42
        Your shortcuts 44
        The littlest links 46
    Right On 46
        Sponsored 47
        Your Pages 47
        Birthdays..47
        Contacts 47

PART 2: DAY-TO-DAY FACEBOOK 49

CHAPTER 4: **Reading News Feed and Posting** 51
    Your Daily News . . . Feed..52
        Anatomy of a News Feed post..................................53
        Common actions and content...................................54
        Checking out stories 59
        Viewing stories...60
    Interacting with Your News Feed 61
        Liking 61
        Reacting 62
        Liking (or following) Pages 62
        Commenting..63
        Sharing 66
        Saving 67

Adjusting News Feed. 68
    Hiding posts and people. 68
    News Feed preferences 70
Sharing Your Own News. 71
    Status updates. 72
    Figuring out what to say 73
    Beyond the basic status update 74
    Creating a story 78
    Controlling who sees your posts. 80

CHAPTER 5: Timeline: The Story of You 83
Making a First Impression 84
    Changing your cover photo 85
    Editing your profile picture 86
    Adding a frame to your profile picture. 88
    Adding a bio 89
Telling Your Story 90
    Creating posts 90
    Creating life events 93
    Editing posts. 95
    Checking out your intro and more 96
Telling the World about Yourself. 98
    Adding work and education info. 99
    Adding the places you've lived. 101
    Adding contact and basic info 102
    Adding family and relationships 103
    Adding details about yourself 105
    Adding life events 105
Viewing Timeline Tabs 105
Your Friends and Your Timeline 106

CHAPTER 6: Understanding Privacy and Safety 109
Knowing Your Audience 110
Changing Privacy as You Post 113
Understanding Your Timeline Privacy 114
Getting a Privacy Checkup 115
    Who Can See What You Share topic 116
    How to Keep Your Account Secure topic 117
    How People Can Find You on Facebook topic. 118
    Your Data Settings on Facebook topic 119
    Your Ad Preferences on Facebook topic 119
Navigating the Settings Page. 120

General section 121
Security and Login section 122
Your Facebook Information section 124
Privacy section 127
Face Recognition section 129
Profile and Tagging section 129
Public Posts section. 131
Blocking section. 133
Location section. 136
Language and Region section 137
Stories section 138
Notifications section 138
Mobile section 140
Apps and Websites section 140
Instant Games section 141
Business Integrations section 142
Ads section 142
Advertisers section 142
Ad Topics section. 143
Ad Settings section 143
Ads Payments section. 145
Facebook Pay section 145
Support Inbox section. 145
Videos section 146
Understanding Privacy Shortcuts 146
Checking out Facebook's privacy tools. .149
Remembering that it takes a village to raise a Facebook 152
Peeking Behind the Scenes 153
Protecting minors 154
Preventing spam and viruses 154
Preventing phishing 154
One Final Call to Use Your Common Sense. .155

CHAPTER 7: **Facebook on the Go** 157
The Facebook App. 158
Layout and navigation 158
News Feed 164
Reacting to Posts. 165
Commenting on posts 166
Post and News Feed options 166
Posting from the App 168
Photo Posts 170
Taking photos and creating videos to share 175
Creating Facebook stories 177
Viewing and interacting with stories. 180

Checking Out Timelines 181
Using Groups 185
  Viewing an event 185
Facebook Messenger 186
  Navigating Messenger 187
  Viewing and sending messages. 188
  Video calls. 190
The Facebook Family of Mobile Apps. 190
  Instagram 190
  WhatsApp 191
  Messenger Kids 191
  Facebook Local. 191
Facebook on Your Mobile Browser. 191
  Mobile Home 192
  Mobile timelines 193
  Mobile inbox. 194
Facebook Texts 194
  Mobile settings. 196
  Mobile notifications. 196

PART 3: CONNECTING WITH FRIENDS. 197

CHAPTER 8: **Finding Facebook Friends** 199
What Is a Facebook Friend? 200
Adding Friends. 201
  Sending friend requests 201
  Accepting friend requests 202
  Choosing your friends wisely. 203
Finding Your Friends on Facebook 203
  Checking out people you may know. 203
  Browsing friends' friends 204
  Using the search box 205
Managing How You Interact with Friends 207
  News Feed preferences 207
  Following. 208
  Unfriending. 209

CHAPTER 9: **Just between You and Me:**
            Facebook Messenger. 211
Sending a Message 212
  Sending a group message 213
  Sending a link. 214
  Sending a photo. 214
  Sending a sticker 214
  Sending a GIF 215

Sending an emoji 215
Sending payment 216
Sending an attachment 217
Sending an instant emoji 217
Starting a video or voice call 217
Managing Messages 218
Checking Out the Chat List 221
Navigating Messenger 222
Message requests 224
Conversations in the inbox 225
Messenger settings 226
Getting into Rooms 226
Messaging on the Go Using the Messenger App 228
Messenger Kids 229

CHAPTER 10: **Sharing with Facebook Groups** 231
Evaluating a Group 232
Sharing with a Group 235
Using the share box 235
Creating events 238
Using files and docs 239
Reading and commenting on posts 242
Group Dynamics 243
Controlling notifications 243
Searching a group 244
Adding friends to a group 245
Creating Your Own Groups 245
Adding detail to your group 247
Deciding a group type 248
Being a Group Administrator 248
Scheduling posts 249
Pinning announcements 249
Managing a group 250
Adjusting group settings 252
Interpreting insights 253
Editing members 253
Reporting offensive groups and posts 255

PART 4: GETTING THE MOST FROM FACEBOOK 257

CHAPTER 11: **Filling Facebook with Photos and Videos** 259
Viewing Photos from Friends 259
Photos in News Feed 260
Photo viewer 261
The album view 262

Viewing photos on your mobile device 263
Viewing tagged photos and videos of yourself 264
Adding Photos to Facebook. 265
Uploading photos 265
Creating an album. 271
Editing and Tagging Photos 274
Editing albums 274
Editing a photo. 275
Automatic albums 278
Working with Video. 278
Viewing videos 279
Adding a video from your computer 280
Adding a video from the Facebook app .280
Live video 281
Discovering Privacy 282
Photo and video privacy. 282
Privacy settings for photos and videos of yourself. 283

CHAPTER 12: **Buying, Selling, and Fundraising** 285
Getting the Most Out of Marketplace. 286
Browsing and buying in Marketplace 286
Live shopping. 289
Selling your stuff on Marketplace 290
Posting jobs on Marketplace 292
Marketplace inbox. 294
Using Marketplace on your phone 294
Belonging to Buy/Sell Groups 297
Browsing and buying in a Buy/Sell group 297
Selling items in a Buy/Sell group. 298
Using Buy/Sell groups on your phone 299
Fundraising for Causes. 300
Donating to a fundraiser 300
Facebook Pay. 301
Creating your own fundraiser 302
Promoting and managing your fundraiser 304

CHAPTER 13: **Scheduling Your Life with Events** 307
You're Invited! 307
Public Events 310
Viewing Events. 311
Creating an Event 312
Inviting guests 314

Managing Your Event 315
    Editing your event's info 315
    Canceling the event. 316
    Messaging your event's guests 316
    Removing guests 317

CHAPTER 14: **Creating a Page for Promotion** 319
Getting to Know Pages 320
    Anatomy of a Page 320
    Connecting and interacting with Pages 322
Creating a Facebook Page 324
    Do I need a Page? 324
    Creating your Page 325
    Getting started. 326
Sharing as a Page 329
    The share box 330
    Creating specialty posts 334
Using Facebook as Your Page 336
    Liking, commenting on, and sharing posts 336
    Liking other Pages 337
Managing a Page 338
    News Feed 338
    Inbox 339
    Business App Store 342
    Resources & Tools. 342
    Manage Jobs. 342
    Notifications 342
    Insights: Finding out who is using your Page 343
    Publishing tools 345
    Ad Center 346
    Page Quality 346
    Edit Page Info 346
    Page settings 347

CHAPTER 15: **Using Facebook with Games, Websites,
and Apps** 355
Understanding What Apps Need 356
    The basics. 356
    The slightly less basics 357
    Permission to act. 358
Games on Facebook 358
    Playing instant games on Facebook 358
    Playing web games on Facebook 360

Inviting and notifying 361
Posting 361
Watching game videos 362
Viewing your gaming activity 363
Keeping your games close 363
Using Facebook Outside Facebook 364
Mobile Apps and Facebook 366
Managing Your Games, Websites, and Apps 367
Adjusting your app permissions 369
Making additional app settings 369
Learning more 369
Removing apps 369
Adjusting your preferences 370
Controlling what you see from friends 371
Reporting offensive apps 371

PART 5: THE PART OF TENS 373

CHAPTER 16: **Ten Ways to Make the Most of Your Facebook Content** 375
Remembering the Past 375
Scrapbooking Baby Photos (Mobile Only) 376
Framing Your Profile Picture 377
Adding Dimension 377
Giving Your Photos Some Flair 378
Reviewing the Last Year (or Years) 379
Making Your Status Stand Out 379
Tagging It All 380
Using Stickers or GIFs in Your Messages 380
Friend-a-versaries 381

CHAPTER 17: **Ten Ways to Be Politically Active on Facebook** 383
Familiarize Yourself with the Voting Information Center 383
Share Your Voting Status 384
Spend Time in the Town Hall 385
Join Relevant Groups 386
Fundraise for a Cause 386
Organize a Digital Grassroots Campaign 386
Fact-Check Your Sources 387
Beware of Click-Bait 387
Don't Feed the Trolls 388
Mute What Drives You Crazy 388

CHAPTER 18: **Ten Frequently Asked Questions** .389

Do People Know When I Look at Their Timelines? 390

I Friended Too Many People and Now I Don't
Like Sharing Stuff — What Can I Do? 390

Facebook Looks Different — Can I Change It Back? 391

I Have a Problem with My Account — Can You Help Me? 392

What Do I Do with Friend Requests I Don't Want to Accept? 392

Why Can't I Find My Friend? 393

Will Facebook Start Charging Me to Use the Site? 394

How Do I Convince My Friends to Join? 394

What If I Don't Want Everyone Knowing My Business? 395

Does Facebook Have a Feature That Lets Me Lock Myself
Out for a Few Hours? 396

INDEX 397

Introduction

Facebook connects you with the people you know and care about. It enables you to communicate, stay up-to-date, and keep in touch with friends and family anywhere. It facilitates your relationships online to help enhance them in person. Specifically, Facebook connects you with the *people* you know around *content* that's important to you. Whether you're the type to take photos or look at them, or write about your life or read about your friends' lives, Facebook is designed to enable you to succeed. Maybe you like to share websites and news, play games, plan events, organize groups of people, or promote your business. Whatever you prefer, Facebook has you covered.

Facebook offers you control. Communication and information sharing are powerful only when you can do what you want within your comfort zone. Nearly every piece of information and means of connecting on Facebook come with full privacy controls, allowing you to share and communicate exactly how — and with whom — you desire.

Facebook welcomes everyone: students and professionals, grandchildren (as long as they're at least age 13), parents, grandparents, busy people, celebrities, distant friends, and roommates. No matter who you are, using Facebook can add value to your life.

About This Book

Part 1 teaches you the basics to get you up and running on Facebook. This information is more than enough for you to discover Facebook's value. Part 2 teaches you about using Facebook — the sorts of things millions of people log in and do every day. Part 3 explains how to find friends and all the ways you can interact with them. Part 4 explores some of the special ways you might find yourself using Facebook. Finally, Part 5 explores the creative, diverse, touching, and even frustrating ways people have welcomed Facebook into their lives.

Here are some of the things you can do with this book:

>> **Find out how to represent yourself online.** Facebook lets you create a profile (called a timeline) that you can share with friends, coworkers, and people you have yet to meet.

>> **Connect and share with people you know.** Whether you're seeking close friends or long-lost ones, family members, business contacts, teammates, businesses, or celebrities, Facebook keeps you connected. Never say, "Goodbye" again . . . unless you want to.

>> **Discover how the online tools of Facebook can help enhance your relationships offline.** Photo sharing, group organization, event planning, and messaging tools all enable you to maintain an active social life in the real world.

>> **Take Facebook with you when you're not at your computer.** Facebook's mobile tools are designed to make it easy to use Facebook wherever you are.

>> **Bring your Facebook connections to the rest of the web.** Many websites, games, apps, and services on the Internet can work with your Facebook information to deliver you a better experience.

>> **Promote a business, a fundraiser, or yourself to the people who can bring you success.** Engaging with people on Facebook can help ensure that your message is heard.

Foolish Assumptions

In this book, we make the following assumptions:

>> You're at least 13 years of age.

>> You have some access to the Internet, an email address, and a web browser that is not Internet Explorer (Safari, Chrome, Firefox, and so on are all good).

>> There are people in your life with whom you communicate.

We state our opinions throughout this book. Although we've worked for Facebook in the past, the opinions expressed here represent our own perspectives, not that of Facebook. We were avid Facebook users long before either one of us worked for Facebook.

Icons Used in This Book

What's a *Dummies* book without icons pointing you in the direction of great information that's sure to help you along your way? In this section, we briefly describe each icon used in this book.

TIP

The Tip icon points out helpful information that's likely to improve your experience.

REMEMBER

The Remember icon marks an interesting and useful fact — something you may want to use later.

TECHNICAL
STUFF

The Technical Stuff icon indicates interesting and probably unnecessary information that might prove useful at some later point.

WARNING

The Warning icon highlights lurking danger. With this icon, we're telling you to pay attention and proceed with caution.

Beyond the Book

In addition to what you're reading right now, this product also comes with a free access-anywhere cheat sheet that helps you build your friends list, communicate with your friends in the many ways available on Facebook, and stay on top of important Facebook dates, such as friends' birthdays and events. To get the cheat sheet, simply go to www.dummies.com and enter *Facebook For Dummies Cheat Sheet* in the search box.

Where to Go from Here

Now that we're properly introduced, let's get started with Facebook. If you're a new user, we recommend starting with Part 1, where we explain the different ways people use Facebook, how to set up an account, and how to navigate around the site. If you already have those basics down, you can head over to Part 2, which walks you through the Facebook features you'll likely interact with every day — News Feed, timelines, privacy, and using Facebook on your mobile phone.

If you're feeling a little isolated on Facebook, Part 3 is where you should start. The information there will help you find friends, connect to other people through groups, and use Facebook Messenger to stay in touch. In Part 4, we dive deeper into some of the features that help enhance your Facebook experience such as photos, games, Pages, and fundraising for causes. And in Part 5, our many accumulated years of Facebook experience can answer some of your FAQs and provide tips and tricks for using Facebook in different ways.

1
Getting Started

IN THIS PART . . .

What you can and can't do on Facebook

Signing up and getting confirmed

Looking around and navigating Facebook

IN THIS CHAPTER

» **Understanding Facebook**

» **Knowing what you can and can't do on Facebook**

» **Finding out how Facebook is different from other social sites**

» **Seeing how different people use Facebook differently**

Chapter **1**

The Many Faces of Facebook

Think about the people you interacted with throughout the past day. In the morning, you may have gone outside to get the paper and chatted with a neighbor. You may have asked your kids what time they'd be home and negotiated with your partner about whose turn it is to cook dinner. Perhaps you spent the day at the office, chatting, joking, and (heaven forbid) getting things done with your co-workers. In the midst of it all, you may have sent an email to all the people in your book club, asking them what book should be next and what date works for the most people. Maybe while you sat on the bus you read the newspaper or called your mom to wish her a happy birthday or searched on your phone for a good restaurant to go to for drinks with friends. This is your world, as it revolves around you.

Each of us has our own version of the world, and as we interact with each other, those worlds intertwine, interplay, and interlock. Maybe your best friend from college was the one to introduce you to the book club, and then someone from the book club recommended a good restaurant. This network of people you interact with — your friends, acquaintances, and loved ones — exists online. Facebook is the online representation of the web of connections between people in the real world.

Now, you may be asking, if this network exists in the real world, why do I need it online, too? Good question (gold stars all around). The answer is that having this network online facilitates and improves all your social relationships. In other words, Facebook makes your life easier and your friendships better. It can help with practical things such as remembering a friend's birthday or coordinating a party. It can help also with the more abstract aspects of relationships, things like staying close with family you aren't physically near or talking about your day with friends.

Getting set up and familiar with Facebook does take a little work (which you know, or else you wouldn't be starting on this book-length journey). It may feel a little overwhelming at times, but the reward is worth it — we promise you.

So What Is Facebook, Exactly?

"Yes," you're saying. "I know Facebook is going to help me stay in touch with my friends and communicate with the people in my life, but what *is* it?"

Well, at its most basic, Facebook is a website. You'll find it through a web browser such as Safari, Google Chrome, or Firefox, the same way you might navigate to a search engine like Google or to an airline's website to book tickets. (You can also access it using an app on your smartphone or tablet, but more on Facebook Mobile in Chapter 7.) Figure 1-1 shows what you will probably see when you navigate to www.facebook.com.

FIGURE 1-1:
Welcome to
Facebook.

If you're already a Facebook user and choose to stay logged in to your computer, www.facebook.com will likely look more like Figure 1-2, which shows an example of your News Feed and Home page.

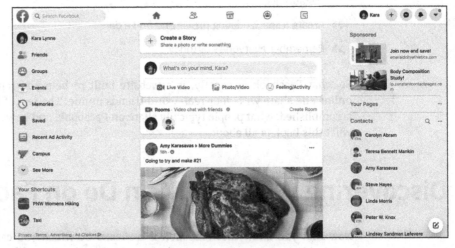

FIGURE 1-2:
Welcome back to Facebook, old friend.

Facebook is a website where you go to connect and share with friends. And just as there are a lot of different ways you interact with friends in the real world, there are a lot of ways to do so on Facebook. For example, you may go to Facebook to

>> Check out what your friends are up to today.

>> Read articles and other news that your friends have posted.

>> Tell your friends and family about your recent successes, show them your photos, or let them know you're thinking of them.

>> Share a tidbit from your day, something you've been thinking about, or an article you found interesting.

>> Show off the pictures from your latest vacation.

>> Share a live video of a concert, an event, or whatever is going on right now.

>> Make a contact in a city you're moving to or at a company where you're applying for a job.

>> Plan an event.

>> Get in touch with an old friend.

>> Garner support for a cause.

>> Buy or sell used items.

>> Get recommendations from friends for movies, books, music, and restaurants.

>> Create a special dating profile and find a date.

>> Remember everyone's birthday.

So what Facebook *is*, exactly, is a website built to help you represent yourself online and share with your real-world friends online. The rest of it — how that's accomplished, what people typically share on Facebook, and how it all works — is what this book is all about.

Discovering What You Can Do on Facebook

Now that you know that Facebook is a means by which you can connect with people who matter to you, your next question may be, "How?" Another gold star for you! In the next few sections, you receive an overview.

Connecting with friends

As soon as you sign up for Facebook, you'll start seeing prompts to "Add Friends." *Friendships* are the digital connections between you and your real-world friends and acquaintances. On Facebook, *friending* just means establishing a virtual connection. Friending people enables you to communicate and share with them more easily. Friends are basically the reason Facebook can be so powerful and useful to people. Facebook offers the following tools to help you find your friends:

>> **People You May Know:** Displays the names and pictures of people you likely know. These people are selected for you based on commonalities such as where you live or work or how many friends you have in common.

>> **Search:** Helps you find the people in your life by name. Chances are they are already using Facebook.

After you establish a few friendships on Facebook, use those friendships to find other people you know by searching through their connections for familiar names. Chapter 8 explains how to find people you know on Facebook.

Discovering what's going on with your friends

Whenever you log into Facebook, you'll see your News Feed. *News Feed* is the constantly updating list of stories by and about your friends. In less vague terms, every time one of your friends adds something to Facebook — a photo, a post about her day, a link to an article he liked — it creates a story that may appear when you log in. In this way, your News Feed becomes an ongoing update about your friends. News Feed is how you know when your friends have become engaged, moved, or had a baby. It's how you know who had a funny thought while waiting for coffee, and whose kid just said something bizarre and profound. It's how you know that there was a tiny earthquake in your area ten minutes ago (don't worry, everyone's fine) and that people were disappointed by the way your city's basketball team played over the weekend. You can see a snippet of a News Feed in Figure 1-2. Chapter 4 provides much more detail about News Feed.

Establishing a timeline

When you sign up for Facebook, one of the first things you do is establish your profile, or timeline. Facebook (and your authors) use these terms interchangeably. On Facebook, a *profile* is much much more than an at-a-glance bio; it updates every time you add something to Facebook, creating an ongoing history of your life on Facebook. When you (or your friends) are feeling nostalgic, you can explore your history the same way you might flip through an old photo album.

At first, the thought of putting a photo album of your entire life online may feel scary or daunting. After all, that stuff is personal. But one of the things you'll discover about Facebook is that it's a place to be personal. The people who will see your timeline are, for the most part, the people you'd show a photo album to in real life. They are your friends and family members.

WARNING

That "for the most part" is an important part of Facebook, too. You'll encounter other people on Facebook, including potential employers or professional contacts, more distant friends, and casual acquaintances. This distinction — between your close friends and everyone else — is an important one to be aware of.

The timeline, which is shown in Figure 1-3, is set up with all kinds of privacy controls to specify *whom* you want to see *which* information. The safest rule here is to share on your timeline any piece of information you'd share with someone in real life. The corollary applies, too: Don't share on your timeline any information that you wouldn't share with someone in real life.

Chapter 5 provides lots of detail about the timeline and what you might choose to share there. For now, think of it as a personal web page that helps you share with your friends on Facebook.

Communicating with Facebook friends

As Facebook grows, it becomes more likely that anyone with whom you're trying to communicate can be reached. Chances are you'll be able to find that person you just met at a dinner party, an old professor from college, or the childhood friend you've been meaning to catch up with. Digging up a person's contact information could require calls to mutual friends, a trip to the white pages (provided you know enough about that person to identify the right contact information), or an email sent to a potentially outdated email address. Facebook streamlines finding and contacting people in one place. If the friend you're reaching out to is active on Facebook, no matter where she lives or how many times she's changed her email address, you can reach each other.

And Facebook isn't just about looking up old friends to say hi. Its messaging system is designed to make it easy to dash off a quick note to friends and get their reply just as fast. The comments people leave on each other's photos, status updates, and posts are real conversations that you will find yourself taking part in.

Sharing your thoughts

You have something to say. We can just tell by the look on your face. Maybe you're proud of the home team, or you're excited for Friday, or you can't believe what you saw on the way to work this morning. All day long, things are happening to all of us that make us just want to turn to our friends and say, "You know what? . . . That's what." Facebook gives you the stage and an eager audience. Chapter 4 shows how you can make short or long posts about the things happening around you and how to distribute those posts easily to your friends.

Sharing your pictures and videos

Since the invention of the modern-day camera, people have been all too eager to yell, "Cheese!" Photographs can make great tour guides on trips down memory lane — but only if you remember to develop, upload, or scrapbook them. Many memories fade when the smiling faces are stuffed into an old shoe box, remain on undeveloped rolls of film, or are left to molder in obscurity on your phone's camera roll.

Facebook offers three great incentives for uploading, organizing, and editing your photos and videos:

» **Facebook provides one easy-to-access location for all your photos and videos.** Directing any interested person to your Facebook timeline is easier than emailing pictures individually, sending a complicated link to a photo site, or waiting until the family reunion to show off the my-how-the-kids-have-grown pics. You can share videos alongside your photos, so people can really get a feel for all parts of your vacation.

» **Every photo and video you upload can be linked to the timelines of the people in the photo or video.** For example, suppose you upload pictures of you and your sister and link them to her timeline. On Facebook, this is called *tagging* someone. Whenever someone visits your sister's timeline, he sees those pictures; he doesn't even have to know you. This feature is great because it introduces longevity to photos. As long as people are visiting your sister's timeline, they can see those pictures. Photo albums no longer have to be something people look at right after the event and maybe again years later.

WARNING

Friends may have certain settings that prevent you from tagging them in photos. In general, people leave this feature turned on, but if you're trying to tag someone and can't, this might be why.

» **Facebook gives you the power to control exactly who has access to your photos and videos.** Every time you upload a photo or create a new photo album on Facebook, you can decide whether who you want to see it: everyone on Facebook, just your friends, or a subset of your friends based on your comfort level. You may choose to show your wedding photos to all your friends, but those of the honeymoon to only a few friends. This control enables you to tailor your audience to those friends who might be most interested. All your friends might enjoy your baby photos, but maybe only your co-workers will care about photos from the recent company party.

Chapter 11 shows how to share your photos and videos.

Planning events

The Facebook *Events* feature is just what it sounds like: a system for creating events, inviting people to them, sending out messages about them, and so on. Your friends and other guests RSVP to events, which allows the event organizers to plan accordingly and allows attendees to receive event reminders. Facebook Events can be used for something as small as a lunch date or as big as a march on Washington, D.C. Sometimes events are abstract rather than physical. For example, someone could create an event for Ride Your Bike to Work Day and hope the invitation spreads far and wide (through friends and friends of friends) to promote awareness. You can use Events to plan barbecues for friends as well as to put together a large reading series. Chapter 13 covers Events in detail.

Joining and creating groups

Facebook *Groups* are also what they sound like: groups of people organized around a common topic or real-world organization. One group may be intimate, such as five best friends who plan several activities together. Another group could be practical — for example, PTA Members of Denver Schools. Within a group, all members can share relevant information, photos, or discussions. Carolyn's groups include one for each kid's classroom at school, one for her *For Dummies* editorial team to update how the writing is going, one for women who like hiking and camping in the Pacific Northwest, and one Buy Nothing group for passing along used items with others in her neighborhood. Groups are covered in detail in Chapter 10.

Whenever you share content on Facebook, you can choose to share it only with members of a certain group. So if you just had a baby and know how much your family is jonesing for new photos, you can share photos with just your family group without inundating the world at large.

Using Facebook around the Internet

Facebook Photos, Groups, and Events are only a small sampling of how you can use Facebook to connect with the people you know. Throughout this book, you find information about how Facebook interacts with the greater Internet. You might see articles recommended by friends when you go to *The New York Times* website or information about what music your friends like when you use Spotify, an Internet radio website. Chapter 15 explains in detail the games, apps, and websites that you can use with your Facebook information.

Many of these websites and applications have been built by outside developers who don't work for Facebook. They include tools to help you edit your photos; create slideshows; play games with friends across the globe; divvy up bills among people who live or hang out together; and exchange information about good movies, music, books, and restaurants. After you become a little more comfortable with the Facebook basics, you can try some of the thousands of applications and websites whose services allow you to interact with your Facebook friends.

Promoting a business

Every day, you interact with your friends and family. You also interact with other people, places and things: a newspaper or magazine, your favorite coffee shop, a celebrity whose marriage travails you can't help but be fascinated by, a television show that has you on the edge of your seat, or a cause that's near and dear to your heart. All these entities can be represented on Facebook through Pages (with a capital P). These Pages look almost exactly like timelines, just for the not-quite-people among us. Instead of becoming friends with Pages, you can like (or follow) them. So when you like a television show (say, *The Daily Show with Trevor Noah*), you'll start to see updates from that Page (*The Daily Show*) in your News Feed. Liking Pages for businesses or causes helps you stay up-to-date with news from them. Chapter 14 covers the ins and outs of Pages.

Fundraising for a cause

One of the things people often do in the world is try to figure out a way to make it better. Every day, people are working on solving lots of hard problems. Facebook fits into this because it can help you spread the word to friends about the causes you're passionate about. And if your friends care about the same things, they in turn might bring along their friends to create a large group of people willing to help out. In addition to simply passing along information, you can create fundraisers where your friends help you reach a charitable goal. Fundraisers are covered in Chapter 12.

THE BIRTH OF THE 'BOOK

In ye olden days, say, the early 2000s, most college freshmen would receive a thinly bound book containing the names and faces of everyone in their matriculating class. These *face books* were useful for matching names to the students seen around campus or for pointing out particular people to friends. However, these face books had several problems. If someone didn't send in his picture, the books were incomplete. They were outdated by junior year because many people looked drastically different, and the books didn't reflect the students who had transferred in or who were from any other class. Finally, they had little information about each person.

In February 2004, Mark Zuckerberg, a sophomore at Harvard, launched an online "book" to which people could upload their photos and personal information, a service that solved many of these problems. Within a month, more than half of the Harvard undergraduates had signed up.

Zuckerberg was then joined by others to help expand the site into other schools. Carolyn was the first Stanford student to receive an account. During the summer of the same year, Facebook moved to Palo Alto, California, where the site and the company kept growing. By December 2004, the site had grown to one million college students. Every time Facebook opened to a new demographic — high school, then work users, then everyone — the rate at which people joined the site continued to increase.

At the end of 2006, the site had more than 10 million users; 2007 closed out with more than 50 million active users. At the time of this book's publication in 2021, that final count has grown so that now more than two and a half billion people across the globe use Facebook to stay in touch.

Keeping in Mind What You Can't Do on Facebook

Facebook is meant to represent real people and real associations; it's also meant to be safe. Many of the rules of participation on Facebook exist to uphold those two goals.

TECHNICAL STUFF

There are things you can't do on Facebook other than what's listed here. For example, you can't look at the photos of someone who has tight privacy settings; you can't prevent ads from showing up from time to time; you can't spin straw into gold. These rules may change how you use Facebook but probably won't change *whether* you use it. The following four rules are highlighted in this section because if any are a problem for you, you probably won't get to the rest of the book.

You can't lie

Okay, you can, but you shouldn't, especially not about your basic information. Facebook's community standards include a commitment to use an authentic identity, which means Facebook wants you to create only one timeline for yourself. You don't *have* to use your real name, but we recommend that you do. (A few exceptions to this rule include teachers wanting to keep some professional distance from their students by using an alias.) However, if you create multiple accounts or fake accounts, there's a good chance they will be flagged, disabled, and removed from Facebook.

You can't be 12 or younger

Seriously. A U.S. law prohibits minors under the age of 13 from creating an online timeline for themselves. This rule, which Facebook enforces, is in place for the safety of minors. If you or someone you know on Facebook is under 13, deactivate (or make him or her deactivate) the account now. If you're reported to the Facebook Community Operations team and they confirm that you're underage, your account will be disabled.

You can't troll, spam, or harass

On the Internet, *trolling* refers to posting deliberately offensive material to websites to get people upset. *Spamming* refers to sending out bulk promotional messages. When we talk about *harassment*, we mean deliberately tormenting or bothering another person or group of people. If you do any of these things on Facebook, there's a good chance your posts will be removed and your account can be shut down.

Facebook is about real people and real connections. It's one thing to message a mutual friend or the occasional stranger whose timeline implies being open to meeting new people if the two of you have matching interests. However, between Facebook's automatic detection systems and user-generated reports, sending too many unsolicited messages is likely to get your account flagged and disabled.

Similarly, Facebook aims to be a trusted environment for people to exchange ideas and information. If people deliberately disturb the peace with pornographic, hateful, or bullying content, that trust is pretty much broken. While there are many places on Facebook where you can find spirited public discussion of controversial topics, Facebook does respond to reports of offensive material and will take down anything it deems hate speech. (The definition of hate speech is a notoriously difficult needle to thread, so a common complaint against Facebook is that it allows too much hateful material to stay up for too long.)

If you see trolling, spam, harassment, or hate speech taking place, you can report the content or person to Facebook (you see how to report a photo, for example, in Chapter 11). Its Community Operations team will investigate the report. If you're getting warnings about things like spamming, chances are you just need to tweak *how* you're using Facebook. For example, you may need to create a Page instead of using your personal account for mass messaging. You see how to promote your business (or yourself) in Chapter 14.

You can't upload illegal content

Facebook users live in virtually every country in the world, so Facebook is often obligated to respect the local laws for its users. Respecting these laws is something Facebook must do regardless of its own position on pornography (where minors can see it), copyrighted material, hate speech, depictions of crimes, and other offensive content. Doing so is also in line with Facebook's value of being a trusted place for people 13 and older.

Realizing How Facebook Is Different from Other Social Sites

Lots of social sites besides Facebook try to help people connect. Some popular sites are Twitter, LinkedIn, Instagram, Tumblr, Snapchat, and WhatsApp. We start with the biggest reason Facebook is different. Literally, the biggest: Facebook has over *two billion* users across the world (yes, billion with a *b*). Other social sites might be popular in one country or another, but Facebook is popular pretty much everywhere.

REMEMBER

If you're going to use only one social networking site, choose Facebook — everyone you want to interact with is already there.

You'll see a lot of similar functionality across different sites: establishing connections, creating timelines, liking content, and so on. However, each site brings a slightly different emphasis in terms of what is important. LinkedIn, for example, helps people with career networking, so it emphasizes professional information and connections. Twitter encourages its members to share short *tweets*, 280-character posts with their connections. Instagram (which is owned by Facebook) encourages its members to share cool photos taken with mobile phones. Snapchat allows people to have video chats with friends while applying silly filters to their image in the video.

You might find some or all these sites useful at different points in time, but Facebook wants to be the one that's always useful in one way or another — so it tries to offer all the functionality we just mentioned . . . and more.

Finding Out How You Can Use Facebook

Now that you know what you can do, generally, on Facebook, it's time to consider some of the specific ways you may find yourself using Facebook in the future. The following list is by no means comprehensive, and we've left out some of the things already mentioned in this chapter (such as sharing photos and events and groups). These are more specific-use cases than an advertisement for Facebook's features.

REMEMBER

Two billion people use Facebook, but not all of them can see your entire timeline. You can share as much or as little with as many or as few people as you desire. Put under lock and key the posts or parts of your timeline you don't want to share with everyone. Chapter 6 goes into much greater detail on how to protect yourself and your information.

Getting information

At some point, you may need to find someone's phone number or connect with a friend of a friend to organize something. Facebook can make these practical tasks easier. If you can search for someone's name, you should be able to find him or her on Facebook and find the information you're looking for.

Keeping up with long-distance friends

These days, families and friends are often spread far and wide across state or country lines. Children go to college; grandparents move to Florida; people move for their job or because they want a change of scenery. These distances make it hard for people to interact in any more significant way than gathering together once a year to share some turkey and pie (pecan, preferably).

Facebook offers a place where you can virtually meet and interact. Create a room where you can hang out virtually with friends; upload photos of the kids for everyone to see; write posts about what everyone is up to. Even the more mundane information about your life ("I'm at jury duty") can make someone across the world feel, just for a second, as though she's sitting next to you and commiserating with you about your jury summons.

Moving to a new city

Landing in a new city with all your worldly belongings and an upside-down map can be hugely intimidating. Having some open arms or at least numbers to call when you arrive can greatly ease the transition. Although you may already know some people who live in your new city, Facebook can help connect with all the old friends and acquaintances you either forgot live there or have moved there since you last heard from them. These people can help you find doctors, apartments, hair stylists, Frisbee leagues, and restaurants.

As you meet more and more new friends, you can connect with them on Facebook. Sooner than you thought possible, when someone posts about construction slowing down his commute, you know exactly the street he means, and you may realize, *I'm home.*

Getting a job

Plenty of people use Facebook as a tool for managing their careers as well as their social lives. If you're considering a job at a company, find people who already work there to get the inside scoop or to land an interview. If you're thinking about moving into a particular industry, browse your friends by past jobs and interests to find someone to connect with. If you go to a conference for professional development, you can keep track of the people you meet there as your Facebook friends. Facebook has a jobs listing portion of the site you can use to browse for jobs in your desired field or area, putting the "networking" in "social networking."

Throwing a reunion

Thanks to life's curveballs, friends at a given time may not be the people in your life at another. The memories of people you consider to be most important fade over the years so that even trying to recall a last name may give you pause. The primary reason for this lapse is a legitimate one: There are only so many hours in a day. While we make new, close friends, others drift away because it's impossible to maintain many intense relationships. Facebook hasn't yet found a way to extend the number of hours in a day, so it can't fix the problem of growing apart. However, Facebook can lessen the finality and inevitability of the distance.

Because Facebook is only about 17 years old (and because you're reading this book), you probably don't have your entire social history mapped out. Some may find it a daunting task to create connections with everyone they've ever known, which we don't recommend. Instead, build your map as you need to or as opportunity presents. Perhaps you want to upload a photo taken from your high school graduation. Search for the people in the photo on Facebook; form the friend connection; and then *tag,* or mark, them as being in the photo. (You can learn about photo tagging in Chapter 11.) Maybe you're thinking about opening a restaurant, and you'd like to contact a friend from college who was headed into the restaurant business after graduation. Perhaps you never told your true feelings to the one who got away. For all these reasons, you may find yourself using the Facebook search box.

Finding a happily ever after

Sometimes after hearing a description of Facebook, people worry that it's some sort of dating site. No, no, no, we always reassure them, it's definitely not a dating site . . . except if you want it to be. Facebook Dating is a separate part of the site that people looking for love can opt into. You set up a separate profile for the dating portion of Facebook, but then Facebook uses information about the sorts of groups you've joined and events you've attended to find matches based on your interests. Everything that happens on Facebook Dating stays inside Facebook Dating — your messages with potential matches, your profile, their profiles, and so on. It might just be where you find the one.

Entertaining yourself and playing games

Look, keeping up with friends is great, but lots of people log in to Facebook simply to be entertained. Facebook uses cues from friends to try and find the videos that are most likely to be of interest to you. Additionally, Facebook produces original content that can be found in a section of the site called Facebook Watch. If you enjoy gaming and watching livestreams of esports, there are also lots of ways to have that itch scratched in the Gaming sections of Facebook. You can learn more about gaming in Chapter 15.

Communicating in times of trouble

It's a sad fact of life that sometimes events happen beyond our control. Disasters great and small befall everyone at one time or another. While Facebook tends to be a place for sharing the good stuff, its tools also work very well to help with some of the logistics of recovering from certain types of disasters. *Safety Check* is a feature that gets turned on in certain geographic regions after natural disasters or security attacks. This feature allows people to easily notify their wider Facebook community that they are okay and can even help them coordinate with the services they might need. Facebook's Groups feature was used to help coordinate civilian boat evacuations after a hurricane flooded Houston, Texas, in 2017. Because people live so much of their lives on Facebook, Facebook winds up being there for both the good and the bad.

Chapter **2**

Adding Your Face

C hapter 1 covers why you might want to join Facebook. In this chapter, you find out how to sign up for Facebook and begin using the site. Keep a few things in mind when you sign up. First, Facebook is exponentially more useful and more fun when you start adding friends. Without friends, it can feel kind of dull. Second, your friends may take a few days to respond to your friend requests, so be patient. Even if your first time on Facebook isn't as exciting as you'd hoped, try again over the following weeks. Third, you can have only one personal account on Facebook. Facebook links accounts to email addresses or mobile numbers, and your email address (or number) can be linked to only one account. This system enforces a world where people are who they say they are on Facebook.

Signing Up for Facebook

Officially, all you need to join Facebook is a valid email address or valid mobile number. When we say *valid email*, we mean that you need to easily access the messages in that account because Facebook emails you a registration confirmation. A *valid mobile number* means a mobile phone number that can send and receive text messages, because Facebook will text you your registration confirmation. Figure 2-1 shows the crucial part of the sign-up page, which you can find by navigating to www.facebook.com.

As you can see, you need to fill out a few things:

>> **First and Last Name:** Facebook is a place based on real identity. Sign up with the name people know you by. We don't recommend signing up with a fake name or alias because that will make it hard for your friends to find you on the site. After you've signed up, you can add nicknames or maiden names to your timeline to make it even easier for friends to find you. But for now, just use your real first and last name.

>> **Mobile Number or Email:** You need to enter your valid email address or mobile phone number here. If you enter your email, you'll need to do so twice to make sure there are no typos.

FIGURE 2-1:
Enter information here to create a Facebook account.

>> **New Password:** As with all passwords, using a combination of letters, numbers, and punctuation marks is a good idea for your Facebook password. It's probably not a good idea to use the same password for every site you join, so we recommend using something unique for Facebook. Facebook requires passwords to be at least six characters.

>> **Birthday:** Enter your date of birth. If you're shy about sharing your birthday, don't worry: You'll be able to hide this information on your timeline later.

>> **Gender (Female, Male, or Custom):** Facebook uses your gender information to construct sentences about you on the site. For example, you might see a News Feed story that reads "Amy updated her profile picture." Your gender options are Female, Male, or Custom. You must choose one. If you choose Custom, you'll see a drop-down menu to select your preferred pronoun. Your options are she/her, he/him, or they/them. Your pronoun will be visible to everyone on Facebook, and Facebook will use it to construct sentences about you. In the text field below the pronoun menu, you can enter your gender separately if it differs from your pronoun or you want to add more context to your gender identity. You can also leave this gender text field blank.

After you fill out this information, click Sign Up (the big green button). Congratulations: You've officially joined Facebook!

REMEMBER

When you click Sign Up, you're agreeing to Facebook's terms of service, data policy, and cookies policy. Most websites have similar terms and policies, but if you're curious about just what Facebook's are, you can click the blue Terms, Data Policy, and Cookies Policy links just above the big green Sign Up button.

Checking Your Inbox

After you sign up for Facebook, you'll immediately receive an email in your inbox asking you to confirm your account. This may be the first of many emails Facebook sends you as it helps you get fully integrated into the Facebook world. Read on to learn how to respond to these emails and why they are important.

Confirmation

Confirmation is Facebook's way of trying to make sure you are really you and that the email address you used to sign up is really yours. When you click the Sign Up button, Facebook sends you an email asking you to confirm your account. In other words, Facebook is double-checking that you are the person who owns your email address.

TIP

If Facebook is asking you to confirm your email but you aren't seeing that email in your inbox, try checking your spam or trash folder. Sometimes Facebook emails can wind up there by accident.

Go to your email, look for the Facebook message, and open it. (It will usually have a subject such as *Welcome to Facebook* or *Facebook Confirmation*.) That email contains a link or button. Your confirmation email may also contain a confirmation code that you will be asked to enter on Facebook's website. Click the link or button, enter the confirmation code if prompted, and you will be confirmed.

Email outreach

After you've confirmed your email address Facebook considers you a full-fledged member of the site. However, it doesn't want you to show up once and leave, so it may email you to remind you that you're now a Facebook user. These outreach emails have various subject lines, ranging from a notice that one of your new Facebook friends has updated his or her status, to a general notice that "You have more friends on Facebook than you think." Clicking the links in these emails will open Facebook in your browser.

If you don't like receiving these emails, you can unsubscribe by clicking the Unsubscribe link at the bottom of any individual email. Facebook opens in your browser and asks if you're sure you want to unsubscribe from this type of email, as shown in Figure 2-2. Click Confirm to make it official.

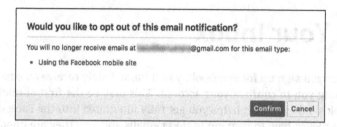

FIGURE 2-2:
Keep your inbox
free of clutter.

>
>
> **WARNING**
>
> When Facebook asks whether you want to unsubscribe from "this type" of email, it is being specific. Email updates about your friends (such as, *Amy added new photos on Facebook*) are a different type than general prompts to find more friends. You may have to click Unsubscribe from more than one email before you stop receiving emails altogether.

Getting Started

Now that your account is confirmed, you're ready to take Facebook by storm. Although you have this book to help guide you through the ins and outs of Facebook, lots of Facebook users do not. (How sad for them!) That's why Facebook provides its users with numbered steps to help start them on the right foot.

In some cases, depending on whether you were invited to join Facebook by a friend or you joined with an email address from your workplace or school, you may see slightly different numbered steps than those detailed in the following sections. Don't worry if this is the case; the same principles apply. This section covers the most basic parts of getting started: finding friends and making yourself recognizable to them by adding a profile picture.

Step 1: Adding a profile picture

Your *Facebook timeline* is the online representation of who you are. Most likely, you have online profiles for various websites. Facebook timelines tend to be a little more comprehensive and dynamic, for reasons that we detail in Chapter 5.

Your profile picture is one of the most important parts of your timeline. It's a good first step to telling your friends all about you. And, significantly, it helps your friends identify you once you start sending friend requests.

You can add a profile picture by uploading a photo from your computer's hard drive.

To add a profile picture from your hard drive, make sure you have a photo you want to use saved somewhere you can find it, and follow these steps:

1. **Click the blue Add Picture button.**

   This opens a window for browsing your computer's hard drive. Use it to navigate to wherever you saved the photo you want to use as a profile picture.

2. **Select your desired photo and click Choose, OK, or Open.**

   This brings you back to where you started, except now there's a preview of your new profile picture. You can choose to keep this photo or upload a different one.

TIP

We talk more about your profile picture and the many ways it's used on Facebook in Chapter 5, but here are a few quick tips on selecting a profile picture:

>> **Make a good first impression.** Your profile picture is how you choose to represent yourself and one of the first ways people interact with your timeline. Most people choose pictures that are flattering or represent what's important to them. Sometimes, profile pictures include other people — friends or significant others. Other times, the location matters. If the first photo you see of someone is at the beach rather than, say, at a party or in an office, you may draw different conclusions about that person. What picture represents you?

>> **Consider who will see your profile picture.** By default, your profile picture appears in search results that are visible to all of Facebook and can even be made available to the larger Internet population. So, generally, people who search for your name can see that picture. Make sure it's something you're comfortable with everyone seeing.

>> **Choose a photo *you* like.** As you use Facebook, you'll wind up seeing your own photo quite often. Small versions appear wherever you make a comment, post something, or are part of a group. So choose a photo you like looking at.

>> **Realize that you're not stuck with the photo.** You can easily change your profile picture at any time. Is it the middle of winter, and that photo of you hiking last summer is just too depressing to look at? No problem; simply edit your profile picture. After you've uploaded your profile picture, you'll also see an option to make the picture temporary for any period of time between 1 hour and forever. If you set it for a length of time with an end date, your profile picture will simply revert to the previous profile picture on that date. See Chapter 5 for details.

Step 2: Finding your friends

Now that your profile picture is in place, it's time to find your friends. Without friends, Facebook can feel a bit like going to an amusement park alone. Sure, the rides were fun and the food was greasy, but no one was there to appreciate it with you.

You have many ways to find friends on Facebook, as you discover in Chapter 8, along with understanding what friendship means on Facebook.

To find your friends, follow these steps:

1. **Enter a name or email in the search field.**

   Facebook might show you a drop-down list with possible matches. You can choose someone from the list or click the Return or Enter key to search for the name you typed in the search field.

2. **Browse the results.**

   Facebook shows you a list of people with exact or similar names to what you typed, as shown in Figure 2-3. To ensure that you've found the correct person, click the name and check out that person's timeline and profile picture.

FIGURE 2-3:
Find your friends
early and often.

3. **Use the filters in the left sidebar.**

   If you're looking for someone with a common name, the list of suggestions might seem endless. Never fear! Facebook gives you some handy-dandy filters in the left sidebar to narrow your search. Your options are to filter by Friends of Friends (particularly useful if you have mutual friends), City, Education, or Work.

4. **Send a friend request.**

   After you have determined that this is your friend, click the Add Friend icon to the right or below the person's name.

5. **Repeat, repeat, repeat.**

   Keep searching and keep friending. We discuss more ways to find friends in Chapter 8, but this is a good way to get started.

TIP

Depending on how you joined Facebook, you may see a few friend suggestions waiting for you below the friend search tool. These are people that Facebook thinks you may know. If you know them, great — go ahead and send a friend request. If not, you can click the Remove button to get rid of the suggestion.

TIP

Sometimes, you might find friends in the search results but you don't see an option to send them a friend request. If this is the case, they might have set their privacy to allow requests only from friends of friends or people who are already Facebook friends with someone that they're Facebook friends with. Try sending them a message (see Chapter 9) to let them know you'd like to be Facebook friends.

Step 3: Getting to know your privacy settings

Click the blue Take a Privacy Tour button to learn more about your privacy settings and options. We won't get into Facebook's many privacy settings here. For the skinny on all things privacy, head to Chapter 6.

REMEMBER

These same steps apply if you sign up for a Facebook account using your mobile phone. The only difference is that you may be able to upload your profile picture and find friends before being prompted to confirm your account.

Introducing Your New Home Page

After you complete the last step in the preceding section, you'll see your Home page in all its glory. The Home page (see Figure 2-4) is what you see each time you log in to Facebook.

FIGURE 2-4:
Home sweet
Home page.

While some parts of the Home page remain the same (such as the top bar and left sidebar), the bulk of what you see is constantly changing. This is because the Home page contains the News Feed, which is essentially a real-time newspaper filled with updates showing you what your friends are posting, sharing, and talking about on Facebook.

At the beginning of this chapter, we point out that Facebook gets exponentially better after you have friends. Until your friends respond to your requests, you may not see much on the Home page except prompts to learn more about Facebook, find more friends, or create your own post. After you add the people you know as friends, take a break. Stretch. Go for a walk. Drink some water. Come back over the next few days to see the interesting photos, status updates, and links your friends are sharing.

TIP

As you navigate Facebook for the first time, you may notice small boxes popping up in different parts of the screen, such as the one in Figure 2-5. Don't ignore these! They are trying to teach you tips and tricks to get you comfortable using Facebook.

Your Cover Photo

Select a photo to appear at the top of your profile. It could be from a recent trip or something you're proud of.

Learn More   Skip   OK

FIGURE 2-5:
Little boxes like this one point out how and why to use different Facebook features.

AM I TOO OLD FOR FACEBOOK?

Are you too old for Facebook? Most emphatically, no. Although how people use the site can be different at different ages, Facebook's utility and nature aren't limited to being useful to only young people.

More and more people in older age demographics are signing up for Facebook every day to keep in touch with friends and family members, share photos, create events, and connect with local organizations. Almost everything we discuss in this book is non-age-specific. Everyone — young or old, in college, working, or retired — has networks of friends and people with whom they interact daily. Facebook tries to map these real-world connections to make it easier for people to share information with their friends. Generally, you should feel confident that you and your friends can connect and use Facebook in a meaningful way.

More than two billion people are using Facebook, and that number isn't made up of "a bunch of kids." Rather, it's a bunch of people from every age group, every country, and every walk of life.

Adding More Friends

As a new user logging into Facebook, Facebook overwhelmingly seems to want you to do one thing — add more friends. You may be seeing the updates of friends in your News Feed, but you are almost certainly seeing previews of people you may know with big buttons prompting you to Add Friend. You may see these previews in your News Feed and even on your timeline.

Facebook finds people to recommend as friends based largely on the people you're already friends with. Each time you see a Person You May Know suggestion, you can choose to ignore the suggestion by clicking the tiny X in the upper-right corner or clicking the Add Friend button.

In general, especially when you're just starting out, we lean towards suggesting that you add everyone you know and care about as a friend. Family, friends, neighbors, coworkers, teammates, classmates — add 'em all. Adding more friends will make your News Feed more interesting, and Facebook will learn over time whom, exactly, you find most interesting.

REMEMBER

On Facebook, friendships are *reciprocal*. You don't officially become Facebook friends with someone until he or she has approved your friend request.

On the flip side, we do not recommend adding people you don't like as friends. Yes, we mean that one person from your last job who was always super nice to you but secretly drove you insane. Or that second cousin twice removed who always asks you inappropriate questions about your love life. Don't feel obligated to click Add Friend simply because you *know* someone.

WARNING

Categories of people you may not want to add as friends are teachers, doctors, students, and other people you know and like but would like to maintain a more professional relationship with. Facebook is often a place where people are casual, let their (virtual) hair down, and don't censor. It can be unsettling or even inappropriate to see your therapist venting about her patients or for your boss to see photos of you relaxing at the beach when you claimed you were home sick.

Chapter 8 provides much more information about the nuances of friendship and how to know whom to add.

Filling out Your Profile Information

Getting your timeline set up is not a requirement for using Facebook. In fact, your timeline is something that gets built up over time (and doing so is covered in Chapter 5), so we won't give you such a Herculean task right away.

However, a few basic pieces of information will help you find your friends on Facebook, as well as help your friends identify you when you send them a friend request. This information consists of your current workplace, current city, any schools you've attended, and your hometown. Especially if you have a common name, this information can really help someone who is regarding a friend request figure out if you are in fact Jane Smith from Portland (whom the person definitely wants to be friends with) or Jane Smith from Seattle (not so much).

To add this basic profile information, follow these steps:

1. **From your Home page, click your name on the left sidebar.**

   This takes you to your profile.

2. **Click the About tab.**

   It is located below your name. Clicking it takes you to the About section of your profile, which is likely empty at this time.

3. **Click Add a Workplace, Add a High School, or Add a College.**

   Clicking one of these options opens an interface for typing the name of your workplace, high school, or college, respectively.

4. **Start typing the name of your workplace or school.**

   Facebook *autocompletes*, or attempts to guess at what you're typing as you type. So, for example, if you start typing *mic,* Facebook will display a list of possible company matches — Microsoft, Michelin, Michael Kors, and so on.

5. **Select your workplace or school when you see it appear on the autocomplete list.**

   If your workplace doesn't appear in the list, simply finish typing its name and press Enter.

6. **(Optional) Add more details about your work or school.**

   You can add information such as your specific job title, major, or graduation year.

7. **Click Save.**

   The blue Save button is at the bottom of the section you're editing.

TIP

To the left of the Save button, a small globe icon and the word *Public* lets you know that after you save this information, it will be publicly available. Anyone can see it. However, clicking Public opens a drop-down menu from which you can choose a smaller audience, such as all your Facebook friends or specific Facebook friends. We talk more about the privacy menu and what each option means in Chapter 6.

To add your hometown and current city from the About section of your profile, click Places Lived on the left side of the page (below Work and Education). You can then click to add your current city and hometown the same way you added your work and education information. You can also edit this information from the Overview section.

Chapter **3**

Finding Your Way Around

One of the best things about Facebook is the number of options available. You can look at photos, chat with friends, message friends, read updates from friends . . . the list goes on and on. What does get a little confusing is that there's no one way to do anything on Facebook. Depending on the page you're on, you'll see slightly different things. And depending on who your friends are, you'll see slightly different things. Using Facebook can't exactly be broken down into ten easy steps.

However, you can learn to recognize the elements that are more constant. Starting from when you log in, you will always start on your Home page, which is where you'll find one of Facebook's most defining features, News Feed. Although your News Feed is always different (more on that later) the Home page has a few constants that are detailed in this chapter. If you ever find yourself lost on Facebook (it happens; trust us), click the Home icon or the Facebook logo (in the top bar of any page) to go to the Home page, where you'll be able to reorient yourself.

Figure 3-1 shows a sample Home page. This chapter details the elements of the Home page that you're likely to see, too: menus, lists, and icons that take you to other parts of the site. Some of these can be found no matter where you are on Facebook, some appear only when you're on your Home page, and some will be there, well, sometimes. Learning about these lists, menus, and icons helps you understand how to find your way around Facebook and enables you to work with some of Facebook's features and options.

FIGURE 3-1:
Your Home page
may look a little
like this.

Checking Out the Top Bar

Given all the exciting things happening on the Home page, you might not notice the top bar at first. It is, after all, just a white bar with some icons, as shown in Figure 3-2. But those icons represent some of the most important places on Facebook.

FIGURE 3-2:
The top bar.

The top bar is similar to the navigational bar in the Facebook app (detailed in Chapter 7), so being comfortable with it here will help you switch between your phone and your computer with ease. Wherever you go on Facebook, the top bar will follow you like a loyal puppy, ready to help you find your way back home.

WARNING

Keep in mind that the examples you see throughout this book might look a little different from your own screen. For example, Page owners might see a flag logo so that they can get to their Pages. Other people might see the Facebook Watch logo so that they can quickly get to video content. Like much of Facebook, what you see depends on the features you use.

The top bar holds 12 icons. We go through them all from left to right:

>> **Facebook logo:** The Facebook logo on the far left serves two purposes. First, it reminds you what website you're using. Second, no matter where you are on Facebook, you can click this icon to return to the Facebook Home page.

>> **Search field or icon:** The text box next to the Facebook logo is the search box. This text area is where you can type any sort of search query. Simply click the search field and start typing what you're looking for. As you do, Facebook opens a list with suggested searches. We talk more about how to find people and other Facebook content later, in the "Search" section of this chapter.

>> **Home icon (house):** This icon is always there to bring you back to the Home page. When in doubt, just go Home and start over. The icons in the center of the top bar (from the home icon through the gaming icon, in this list) are what we think of as *destination icons* — when you click them they take you to a new destination in Facebook.

>> **Friends icon (two people):** This icon takes you to your Friends page, which displays any friend requests you've received. You can come here to look at those potential friends' profiles and respond to their requests. If you don't have any pending friend requests, you can also check out people you may know, which are Facebook's recommendations for potential friends. In general, your friends on Facebook will be the same people you're friends with in real life: friends, family, neighbors, and community members. Friends (and the finding of friends) are covered more in Chapter 8.

>> **Marketplace icon (storefront):** Marketplace is a Facebook's feature that enables people to buy and sell from other Facebook users. Click here to browse items for sale in your area. You can learn more about Marketplace as well as other options for buying and selling on Facebook in Chapter 12.

>> **Groups icon (three people):** Groups on Facebook are online spots for discussing and for organizing all sorts of topics and interests. Click this icon to see what's happening with your groups and find more groups to join. Groups are covered in depth in Chapter 10. If you haven't joined any groups yet, you might not see this icon.

>> **Gaming icon (interlocked tiles):** You can use your Facebook account to participate in all sorts of different online games (covered in more detail in Chapter 15). Clicking this icon will bring you to a gaming dashboard, where you can explore further. If you haven't played any games on Facebook yet, you might not see this icon.

>> **Watch icon (TV):** Facebook Watch is Facebook's product for collecting video content in one place as well as producing original video content. Clicking this icon takes you to the Watch dashboard, where you can check out videos from

Pages you follow, shows you watch or subscribe to, and any other videos you are likely to enjoy. (This icon isn't shown in Figure 3-2.)

» **Your name and profile picture:** If you share a computer with other people, glancing at this photo whenever you use Facebook is an easy way to make sure you're using your Facebook account and not your spouse's or kid's account. Clicking your photo brings you to your timeline.

» **Create icon (+):** Connecting with friends is a two-way street. You want to learn about their lives and they want to learn about yours. The way you "talk" to friends on Facebook is to create content such as posts, stories, and life events. Click the Create icon no matter where you are on Facebook to open a menu where you can take the first step toward creating something new to share:

- *Post:* Posts are the bread and butter of Facebook. These creations can include text, photos, and so much more. We cover the many varieties of posts in Chapter 4.

- *Story:* Stories are similar to posts except they automatically disappear from Facebook after a day. We cover these in Chapter 4.

- *Room:* Rooms are video chat rooms that you can create on Facebook and use to video chat with friends wherever they are, whether or not they use Facebook. You can learn about rooms in Chapter 9.

- *Page:* Chapter 14 teaches you about *Pages,* or special profiles for businesses and organizations. If you're ready to create a Page, start by selecting this option.

- *Ad:* Everyone using Facebook will see ads around the site. A smaller group of people create ads to promote businesses, events, or groups. We don't cover ad creation in this book. (You might want to consult *Facebook Marketing For Dummies,* 6th Edition.) We do, however, cover ad controls when we talk about privacy in Chapter 6.

- *Group:* Chapter 10 explains how to create and use groups to communicate and share with smaller groups of people in Facebook. You can view groups you've already joined by clicking the Groups icon in the top bar, but if you want to get started on a new group, you can create it from this menu.

- *Event:* Chapter 13 covers using Facebook to schedule, plan, and generally coordinate an event happening at a specific time. If you want to get a jump on planning your next shindig, start here.

- *Marketplace Listing:* Marketplace is a Facebook destination for buying and selling stuff. (It's also covered in Chapter 12.) The create menu gives you a shortcut to create a listing for something you want to sell.

- *Fundraiser:* We delve into fundraisers in Chapter 12. Choose this menu option to raise money for causes near and dear to you.

>> **Messenger icon (speech bubble):** People use Messenger to communicate directly to one another (or to each other in groups) on Facebook. This window displays previews from your most recent messages as well as a few options for interacting with your messages. We cover these options — and all of Messenger's many features — in greater depth in Chapter 9:

- *See All in Messenger icon (four arrows):* This option brings you to your inbox, where you can see all your messages and get more options for interacting.

- *New Message icon (pencil on notepad):* This option opens a chat window at the bottom of the screen for sending a message to a friend.

- *Options icon (three dots):* Click this icon to display a menu of options for controlling things such as the sounds Facebook makes when you receive a chat, whether or not people can see if you're active on Facebook currently, and whether people you don't know can send you messages.

- *Search field (Search Messenger box):* You can search here for content in your messages or for messages from certain people.

>> **Notifications icon (bell):** When someone on Facebook has taken an action that involves you or that Facebook thinks you will most likely want to see, you're notified by a red flag here. Notification events include someone tagging you in a photo, someone posting something to a group you belong to, and having a birthday. Click this icon to open a window that displays recent notifications.

Notifications appear as unread **(in bold)** until you click them. You can click the three dots at the top of this window to mark all your notifications as read or adjust your notification settings so you get only the notifications you truly want to see. These three dots appear any time you move your mouse over an individual notification, so you can take specific actions related to that particular item. For example, you may want to stop receiving a certain type of notification or report something to Facebook that you suspect of being spam or inappropriate content. Clicking the three dots displays a menu within the notifications window that will allow you to take those actions.

>> **Account icon (down arrow):** The Account menu appears when you click the down arrow. Here's a rundown of some of the categories you can find on the Account menu:

- *Part:* Clicking this option brings you to your profile.

- *Switch Account:* If you use multiple accounts from the same computer, you can go here to quickly toggle between accounts.

- *Settings and Privacy:* Choosing this option displays the Settings page, where you can change your name, email address, password, or mobile information, or the language you want to use on the site. This is also where you go to find privacy settings (detailed in Chapter 6) and notification settings and to deactivate your account.

- *Help and Support:* This option display the Facebook Help Center, where you can get answers to all the questions that we were unable to answer. While we love to say we have all the answers, sometimes things change or your account has a specific problem that we were not be able to solve.

- *Display and Accessibility:* You can adjust the way Facebook looks on your screen by shifting it to dark mode or changing the font size on your screen. You can also access information and turn on keyboard shortcuts if you only use a keyboard and not a mouse to navigate online.

- *Log Out:* Clicking this option ends your Facebook session. If you share your computer with others, always log out to ensure that another person can't access your Facebook account.

WARNING

If you have the Remember Password option selected when you log in, you won't ever be logged out — even if you close the browser — until you click Log Out. We recommend using the Remember Password option only on a computer you don't share with others.

Search

Search has become an integral part of using the Internet. It's the way we find the info we need — whether that's a business address, a person's contact info, or the year of the great San Francisco earthquake. Facebook's search is also important, though it works a bit differently than a search engine such as Google or Bing.

Most of the time, you will use Search to hop quickly to a friend's timeline or to check out a Page you follow. Simply start typing your friend's name in the search box in the blue bar at the top of the screen. Facebook displays an autocomplete list as you type, showing possible matches as you add more and more letters. When you see the name of the person you're looking for, click it or the picture to go to the person's timeline.

Even though the simplest use of Search is what you'll use most of the time, it's worth noting that Facebook has an incredible database of information that you can search through at any time. You can search through friends' posts, photos, and videos simply by entering a search term in the search box. Given the amount of information you might see in your News Feed on any given day, it can be helpful to search to find that one piece of information you're looking for. (You may know someone posted a link to the best place to pick apples in the fall, but you can't remember when.) The search results page allows you to filter for the latest results, or to look at any of nine categories: posts, people, photos, videos, marketplace, Pages, places, groups, and events.

Viewing Stories and News Feed

This chapter is about navigating Facebook, which is why understanding the top bar is so important. At the same time, the top bar isn't the focus of the Home page. Instead, it's meant to serve as a background to the main events in the center of the page: stories and News Feed. We won't spend a ton of time on these features now because we do a deep dive in Chapter 4, but just for getting your bearings, it's important to understand what's taking up the bulk of the space in the center of your Home page.

Stories

Stories are posts that remain on Facebook for only 24 hours. They often are used to capture the minutiae of a day or an experience — a series of photos or a video, often annotated with text, that tells the story of a person's day. The premise of stories is that a play-by-play of someone's day is fleetingly interesting, but not necessarily something that people want to memorialize forever on their profiles or flood News Feed with multiple posts about.

Stories are previewed just below the top bar in the center of the page. You can click any of the previewed images, as shown in Figure 3-3, to start viewing the associated story, or click your own image to create your own story.

FIGURE 3-3:
Check out stories
from friends
here.

Create a
Story

Your Story

Amy
Karasavas

Carolyn
Abram

WEEK ONE OF REMOTE
SCHOOL IS OVER, AT LAST

News Feed

Imagine that your morning paper, news show, or radio program included an additional section that featured articles solely about the specific people you know. That's a description of News Feed. As long as the people you know are active on

Facebook, you can stay up-to-date with their lives via your News Feed. One friend might post photos from his recent birthday party, another might write a post about her new job, and a third might publish a public event for her upcoming art show. These could all appear as stories in your Facebook News Feed.

A News Feed bonus: You can often use it to stay up-to-date on current events just by seeing what your friends are talking about or by liking (or following) the Pages of real-world news organizations and getting their updates in your News Feed. And when there's unusual weather, Facebook is often where you'll find out about it by way of a flurry of posts.

News Feed is one of the most interesting things about Facebook but also one of the hardest to explain. No matter how we describe seeing a photo of a friend and her new baby pop up in News Feed, it won't be as exciting as when a friend posts those photos. However, we do our best to capture at least a bit of this excitement in Chapter 4.

At the top of News Feed is the share box, shown in Figure 3-4. You use this box to add content to Facebook: status posts, photos, links to articles you find interesting, and so on. These posts also go into your News Feed and may appear in your friends' News Feed. Your friends can then comment, like, and generally interact with you about your post.

FIGURE 3-4:
Share what's on
your mind.

The Left Sidebar

The left side of the Facebook Home page is taken up by a sidebar that contains a list of different destinations in Facebook, as shown in Figure 3-5. The sidebar is divided into two sections: one for options that Facebook chooses, and one for shortcuts specific to how you use Facebook. Finally, at the bottom of the left sidebar are some tiny links that you won't have to use very often.

The top section

One thing you might notice in the top section of the left sidebar is that some of the sections from your top bar are repeated. You can click them in either place. You also might see some items in Figure 3-5 that aren't on your own screen. For example, Voting Information Center tends to show up right before elections but not at other times. If there's a natural disaster happening near you, you might see a Crisis Response entry.

If you click See More (at the bottom of the top section), you can get a sense of how many different destinations, features, and tools Facebook offers. Most people won't need all these, but everyone will wind up loving at least one of them. The following list below is not even close to comprehensive. Instead, it's a mix of features that are popular and those that give you a sense of how many options Facebook provides:

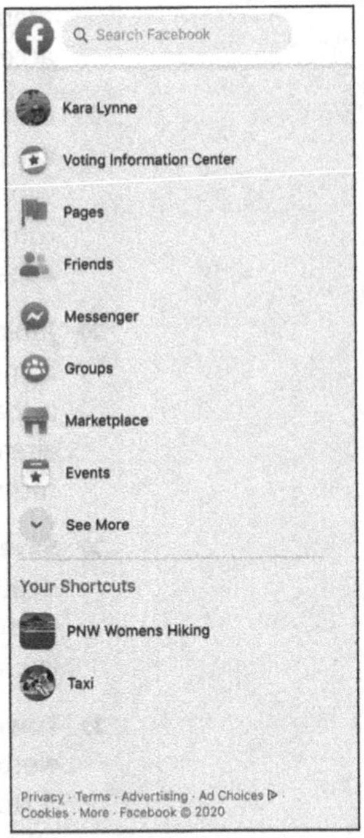

FIGURE 3-5:
A sample sidebar.

>> **Pages:** *Pages* are timelines for everything that's not a regular person. Public figures such as the president or Dwayne "the Rock" Johnson have Pages, as do small businesses, fictional characters, television shows and movies, pets of all hues and stripes, and pretty much everything else you can think of. Clicking the Pages option displays all your Pages as well as tabs where you can view the Pages you have liked (or followed) or find local Pages you may want to like (or follow). Chapter 14 covers creating and managing Pages.

>> **Events:** Facebook's Events feature enables people to easily organize and invite people to an event. You can view events you've been invited to or created by clicking Events in the sidebar. You can learn more about creating, managing, and finding events in Chapter 13.

>> **Fundraisers:** People often use Facebook to promote causes they care about. In response to that, Facebook added the ability to fundraise for a cause. You can fundraise for a personal reason or for a nonprofit. Facebook handles processing your friends' donations and getting them to the organization you select. When you click Fundraisers, you are guided through the process of starting your fundraising and can check out other people's fundraisers. For more information on fundraising with Facebook, see Chapter 12.

>> **Memories:** Memories is a tool that works best after you've been on Facebook for a while. Much like a newspaper callback to notable moments in history on any given day, Memories calls out any notable moments from your own personal history — for example, on this day two years ago you posted a photo of the snake your cat caught in your backyard, or four years ago, you became

friends with someone you later married. Clicking Memories is a great way to access some quick nostalgia.

>> **Games:** Every day, people play various online games through Facebook. You can play games directly with your Facebook friends (most of whom are, as we've mentioned, your real-life friends). Click Games to browse the games you can play and to continue playing games. Interacting with games and other apps is covered in Chapter 15.

>> **Jobs:** You can use Facebook to search for job listings in relevant industries or companies. The benefit of doing this on Facebook instead of another jobs site is that you can try to connect with Facebook friends (or friends of friends) who already work at the companies you're interested in. It puts the *networking* in *social networking*.

>> **Saved:** Often when you're perusing News Feed (which is covered in more depth in Chapter 4), you'll see a link to an article or a video that, for whatever reason, you can't fully appreciate at the moment. You can click to save that link, and then get to it later by clicking Saved in the sidebar.

>> **Town Hall:** Town Hall uses your location information to connect you to your elected officials' Facebook Pages, from your local council-member all the way up to the president. You can follow and send messages to these public officials. You can also turn on voting reminders from Facebook.

>> **Facebook Pay:** Facebook Pay is a method for making secure payments over Facebook. You can connect a credit card or bank account to your Facebook profile and then use that information to make payments to other users, donate to fundraisers, or purchase from online stores operating out of Facebook.

>> **Crisis Response:** When disasters happen, the first thing people want to do is find out if their loved ones are safe. Crisis response enables people watching from the outside get news updates and learn about the ways they can help when something bad is happening. People can volunteer their time, donate money, and search for friends in the affected areas to make sure they're okay.

Your shortcuts

The bottom section of the left sidebar contains shortcuts. Facebook adds items to this section as you use Facebook. For example, if you start to spend a lot of time posting and commenting on a particular group, Facebook will automatically add it to your Shortcuts section. Over time, you may wind up with more shortcuts than easily fit in this space, at which point Facebook will display the shortcuts you click most often; you can hop to the rest by using Search.

In all our time using Facebook, we've never needed to adjust the items in the Shortcuts section of the left sidebar. However, if Facebook's fails you in this regard, you can choose which shortcuts appear in the left sidebar. To change the shortcuts that appear in the Your Shortcuts section, follow these steps:

1. **Hover your mouse cursor over Your Shortcuts in the left sidebar.**

   An Edit button appears to the right.

2. **Click Edit.**

   The Edit Your Shortcuts window appears, as shown in Figure 3-6. Every shortcut appears in this list. Next to each shortcut is a drop-down menu that states whether the shortcut is pinned to the top, sorted automatically, or hidden.

3. **Use the drop-down menus to choose between Pin to Top, Hide, and Sort Automatically.**

   Pinning is the digital equivalent of sticking something at the top of a list and keeping it there. Simply click the menu next to the item you want to change and select whether you want that item pinned, hidden, or sorted automatically. If you hide a shortcut, it will no longer appear in your left sidebar.

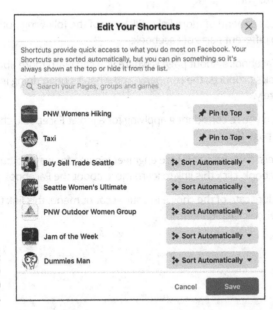

FIGURE 3-6: Use this menu to hide and unpin shortcuts.

The littlest links

At the very bottom of the left sidebar are a handful of important but infrequently needed links. These links all appear in grey text:

- » **Privacy:** View the Facebook data use policy, good for a little light bed-time reading.

- » **Terms:** Check out Facebook's terms and policies, including the Statement of Rights and Responsibilities (which you agreed to when you signed up), the data policy, and its community standards.

- » **Advertising:** Create or manage ads you've posted to Facebook.

- » **Ad Choices:** See the Facebook Help pages, where you can learn more about how Facebook targets ads to you and what you can do to adjust what ads you see.

- » **Cookies:** Sadly, clicking this link doesn't make chocolate chip cookies suddenly appear in your hand. Instead, it brings you to a page that explains how Facebook uses *web cookies,* or stored data on your web browser. Cookies are used on many websites to keep your experience more convenient and to deliver ads to you.

- » **More:** Access a menu of more links. Click any of the following options to navigate to different parts of Facebook:

  - *About:* Facebook's About page is where you can learn more about Facebook's products, the company, and what it's been doing in the news lately.

  - *Careers:* If you're considering applying for a job at Facebook, check out its careers page first.

  - *Developers:* If you're a software engineer looking for ways to build apps that use Facebook, click this link to learn more about the Facebook platform.

  - *Help:* A duplicate of the choice in your Account menu, this link takes you to the Facebook Help Center.

Right On

The right sidebar contains a few sections for other shortcuts that might be useful to you as you use Facebook.

Sponsored

Ads are an inescapable part of using Facebook. You'll see them in various places across the site — in your News Feed, mixed in with stories, and in this section of the right sidebar. You don't need to do anything with this section except let it do its thing, showing you ads. You can learn a bit more about Facebook's ad policies and controls in Chapter 6.

Your Pages

Pages are special profiles created for businesses and public figures, as well as anything else that is not an individual person that requires a Facebook profile. Pages are covered in depth in Chapter 14. If you've already created a Page (or Pages) you can get to them quickly from the Your Pages section of the right sidebar.

Birthdays

If any friends have birthdays coming up, you'll see reminders about them in the right column. One Facebook norm is to write on a friend's timeline to wish him or her a happy birthday. Click your friend's name in the Birthdays section to display a little window that allows you to dash off a quick note.

Contacts

The last section is a list of your friends, sorted by people who are currently active on Facebook. Clicking a name will open a chat window at the bottom of the screen. Much like the top bar, this chat window will stick around no matter where you go in Facebook. Unlike the top bar, however, you can minimize or close chat windows if they are blocking your view. Chats are part of Facebook Messenger; when you close a chat window, you won't lose any of the messages that have been sent back and forth. You can learn all about Messenger in Chapter 9.

2

Day-to-Day Facebook

IN THIS PART . . .

Logging in and reading News Feed

Sharing posts, photos, and links

Managing your privacy

Using Facebook on your mobile phone

Chapter **4**

Reading News Feed and Posting

ometimes after we've explained to people about the basics of Facebook — it's a website that lets you connect and share with your friends — they follow up with an obvious question: "But what do you *do* with it?"

The answer to that question is both simple — you keep up with your friends — and complicated. So many tiny actions and interactions on Facebook add up to a sense of being surrounded by your friends — and that sense of friendship keeps people coming back to Facebook day after day (and, with smartphones, minute after minute).

This book tells you how to do virtually everything you could ever want to do on Facebook, but chances are you won't do some of them — such as create a photo album or plan an event — every single day. As you go about your day online — reading articles, watching videos, shopping, and more — you'll check in on Facebook, find out news big and small from your friends and the Pages you follow, share a few of your thoughts or observations, and go on your way. This chapter covers these basic activities in depth.

The main way people find out news and generally keep in touch on Facebook is through their News Feed, the constantly updating content people post to their timelines. If your friends create *stories*, which are a certain type of post that disappears after 24 hours, you can see what they are up to right this moment, more or less. We talk about News Feed at length in this chapter and about the ways you interact with what you see there. Your News Feed isn't just about reading the posts you see there; it's also about comments and likes.

You can also keep up with friends by sharing your own content — status updates, photos, links to articles you read, and more. You can post your own content and create your own stories from the top of your News Feed. And your friends will, in turn, be able to see, like, and comment on what you posted.

Your Daily News . . . Feed

News Feed is the centerpiece of your Home page. The top bar and the left sidebar are there when you need them but are meant to fade into the background so you can focus on your News Feed.

So what is News Feed? It's the constantly updating content and actions on Facebook of your friends and the Pages you follow — such as writing a status update, sharing a photo or link, or fundraising for charity. See the "Common actions and content" section, later in this chapter, for details. In addition, stories have their own section on top of your News Feed and disappear from Facebook after 24 hours.

Back in the day, a common refrain around the Facebook office was, "News Feed is a robot." More accurately, News Feed is an algorithm. It doesn't show you *everything* from your friends; instead, it shows you things it thinks you will find interesting. News Feed makes its selections based on a complicated calculus of who is posting what and when. News Feed will likely show you more posts from people you interact with more often on Facebook. It also tends to show you big events, such as engagements or new babies from more distant acquaintances.

News Feed also learns the sorts of posts you're likely to click, like, or comment on, and will try to show you more of these. Simply browsing your News Feed and interacting with the things you see helps News Feed improve. You can also manually fine-tune your News Feed as well, as detailed in the "Adjusting News Feed" section.

Because News Feed tries to show you what it thinks you'll find most interesting, posts might not always appear from newest to oldest. Some bias exists toward new posts (especially if you log into Facebook frequently), but you might also see at the top of your News Feed a popular photo that a friend posted several days instead of one from this morning if. Additional likes or comments can also cause a post to reappear in your News Feed even if you've already seen it.

Anatomy of a News Feed post

Figure 4-1 shows a sample News Feed post. In this case, it's a status update from a friend. In the Facebook world, a *status update* refers to any text people post that answers the question "What's on your mind?"

FIGURE 4-1:
Just your average
status update.

Even in this simple example, there are lots of things to pay attention to. However, when you're scanning News Feed, you'll probably first pick up on "who" and "what":

>> **Name and profile picture:** The first part of any post is who it's about or who wrote it. Both the name and picture are links to that person's timeline. In addition, if you hover the mouse cursor over a person's name, you'll see a miniaturized preview of the person's timeline with information about your relationship (you are friends and following her, in most cases) as well as a button you can click to message her.

TIP

Hovering the mouse cursor over any bolded text in a News Feed post generates a preview for a timeline, Page, or interest with specific buttons for adding friends, liking, or following.

>> **Content:** The content section of a News Feed post is the most variable element. You might see a preview of an article, or a video, or a photo album. It could also be a location where someone has *checked in,* or marked her location (such as the Golden Gate Bridge or her local coffee shop) when she posted. The content is the part of the post that is the most important; it's the reason for the post existing. In Figure 4-1, the content is a status update about Kara's houseplants.

After you have the basics and who and what, you can focus on some of the other details to be found in a simple status update:

>> **Feeling/activity info:** Not every status includes this, but Facebook provides a list of emotions and activities that can be appended to any status update or post. In this case, the emoji (and words) depict that Kara was "feeling hopeful" when she wrote this status update.

- » **Tags:** Tags are a way of marking who or what is with you when you post something to Facebook. You might tag a person who is with you when you write a post, or you might tag a TV show you're watching. Tags in posts are displayed as links in blue or bold text. You can hover the mouse cursor over these tags to view more info about that person, Page, place, or thing. In the status update shown in Figure 4-1, Kara Lynne tagged Amy.

- » **Timestamp:** The little grey text near the profile photo in the post tells you how long ago this post was added.

- » **Privacy info:** The grey icon next to the timestamp represents the privacy of that post. Hover the mouse cursor over the icon to see who else can see the post. Usually posts are visible either to everyone (Public) or just to that person's friends.

- » **Like, Comment, and (Share):** These links allow you to interact with your friends about the content they've posted. In addition, you can see how many people have already liked a post, and you can see any comments that have been made below the post itself. You might also see a text box next to your profile picture prompting you to "Write a comment . . .." Commenting, liking, and sharing are covered in more detail in the "Interacting with Your News Feed" section, later in this chapter.

Common actions and content

News Feed is made up of all sorts of posts. Although the basic anatomy is the same, here are some of the common post types you might encounter:

- » **Status updates:** A status update post appears in Figure 4-1. Status updates are the short little posts that your friends make about what's going on in their lives.

- » **Links:** Figure 4-2 shows a post sharing a link. A link from a friend is one of the chief ways many people get their news. Click the link (or the article's title) to go to the article.

- » **Photos and videos:** Figure 4-3 shows a photo album post. When people add photos or are tagged in photos, Facebook creates this type of post, with information about who was tagged and a sample of the photos that were added. You can include videos in photo albums. Videos will autoplay with the volume off as you view them in your News Feed. Click the volume icon (the little loudspeaker) to turn on the sound. Click the photos or videos to see bigger versions, browse albums, and watch more videos.

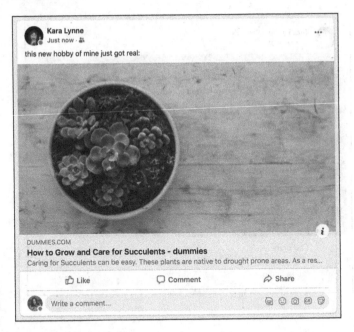

FIGURE 4-2:
Viewing a shared link from a friend.

FIGURE 4-3:
Checking out a friend's photo post.

» **Live videos:** Figure 4-4 shows a live video post. Live videos are just what they sound like — videos streamed live from wherever the poster is. You might see a friend streaming live video from the tide pools she's exploring or a celebrity at a red-carpet event. Many news organizations share live video feeds of formal press conferences, as well as more casual live videos of behind-the-scenes footage. Hover the mouse cursor over the video to display the video's progress bar and other controls. Click the volume icon to turn on the sound and tune in.

» **Group and event posts:** When people post to a group or an event you're a member of, it may show up in your News Feed. These posts look like a standard post, except the top the post shows both the poster's name and the name of the group she's posting to. Figure 4-5 shows Kara Lynne posting to PNW Womens Hiking, a group she is a member of.

» **Life events:** People can create life events from their timelines. These events can be for an event as small as buying a new pair of sunglasses, but more often people use them to mark big moments: a wedding, a baby, moving, getting a pet, buying a house, and other major developments in one's life. Figure 4-6 shows a life event post.

» **Check-ins:** A *check-in* is a way of marking where you are. A post about a check-in show a map of its location, as well as a preview of information about the location where that person checked in. It also may show a list of other friends who have checked in there in the past.

FIGURE 4-5:
A post to a group — with a background!

FIGURE 4-6:
Celebrate your achievements with a life event.

» **Tags:** Posts about tags let you know what photos or posts your friends have recently been tagged in and include a preview of the photos with your friends in them. Because of the way tags work, you may be seeing photos or posts that were added by someone who is not your friend.

» **Memories:** You see two types of memory posts on Facebook. One is basically friends' posts from years ago, which Facebook has shown them and they've chosen to repost. They might add a comment or some context: "Never forget that seven years ago today I ran out of milk." The other type of memory post is a video automatically generated by Facebook to commemorate certain milestones, such as anniversaries and *friendiversaries* (the day two people became friends on Facebook). Memory posts are great for a quick dose of nostalgia.

» **Facebook Watch videos:** Facebook Watch is Facebook's service for professional video content. Facebook produces original television shows as well as serving up videos from other producers such as Netflix and Hulu, or from Pages you follow from your favorite shows and movies. Videos from Facebook Watch automatically begin playing (without sound) when you reach them in your News Feed. To listen to them, click the volume icon. To watch videos in their entirety, click the video. You'll be taken to the Facebook Watch page, a destination in Facebook that compiles all video content in one place for easy sorting, searching, and viewing.

» **Likes and comments:** Posts about likes and comments let you know what Pages, posts, or articles your friends have liked or commented on recently. Usually whatever your friend has interacted with is shown and linked so that you can check it out for yourself.

» **Changed cover and profile pictures:** These posts often look similar to a regular photo post. Click through to look at the new photos on your friends' timelines in their full-sized glory.

» **Events:** Posts about events (usually letting you know which friends have RSVP'd "yes" to an event) include a link to the event, so if you're looking for someplace to go, you can say "yes," too. Only public events show up here, so if you've added a private event, don't worry about people who weren't invited seeing it in their News Feed.

» **Recommendation requests:** Sometimes friends may be looking for help creating a vacation itinerary or finding a good place to buy new soccer cleats. They can ask their wise Facebook friends for help answering these sorts of questions by requesting recommendations. Recommendation request posts usually include location information and a map (like check-in posts) so that you can quickly figure out if you'll be able to help.

» **Fundraisers:** Fundraisers are ways for people to raise money toward a goal. They might be associated with a nonprofit or rounding up cash for a personal goal or on behalf of a friend. When you see posts about your friends' fundraisers, you get some information about the organization or cause, as well as links to learn more or donate. Fundraisers are covered in greater detail in Chapter 12.

» **Read/watch/listen:** Certain services and websites, such as the book-reading site Goodreads, may be allowed to automatically post specific actions people take on their site to Facebook. See Chapter 15 for more information about how these applications work.

» **Apps:** Apps are services and features created by other developers that hook into Facebook and may be allowed to post on behalf of your friends. For example, a game might post when your friend reaches a new level or a workout app might post when your friend has completed a run. You can learn more about apps and games in Chapter 15.

» **Sponsored and suggested:** Suggested and sponsored posts are ads. Ads are what keep Facebook free to use, so there's no way to remove them. These ads aim to be relevant to you and your life and may even help you find Pages or services you find interesting.

Checking out stories

Stories created by your friends appear at the top of your News Feed, directly below the top bar. *Stories* are the most recent content created by your friends. Stories are a type of post, in many ways — they're created by your friends and are often a mix of photos, videos, and text. Unlike standard posts, however, which stick around on people's timelines indefinitely, stories disappear from Facebook after 24 hours, and when you view friends' stories, you see each friend's story in the order they posted it.

For example, let's say you have two friends, David and Alexis. David posted a few times in one day, each time posting a photo of an item for sale in his store. Alexis, instead, makes a story. Each time she adds to the story, she adds another photo of the event she has planned. The photos that David has posted may (or may not) appear in your News Feed over the next several days, depending on the activity of your other friends, how much attention any particular photo has received, and so on. You might see one photo, all of them, or none in your News Feed. The story Alexis has created, on the other hand, will appear as a preview in the Stories section of News Feed for 24 hours. When you click that preview, you will see each photo Alexis has posted, one after another, until you've seen them all. After you've seen all of Alexis's story, you can move on to viewing your next friend's story.

Viewing stories

When you click a story preview at the top of News Feed, Facebook opens Story Viewer, shown in Figure 4-7.

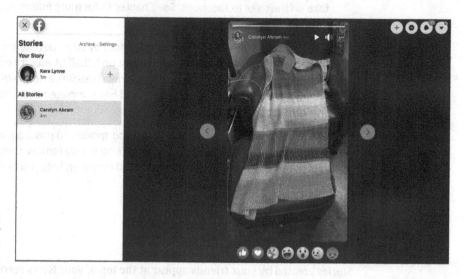

FIGURE 4-7:
Flip through your
friends' stories
here.

Story Viewer has two main areas, the left side, which is white, and the right side, where the story is shown against a black background. On the left side you can see a list of friends (and Pages) that have stories you can view. The person whose story you're currently viewing is highlighted in grey. Below the person's name is some info on how many new posts have been added to the story (if any have), and how long ago the most recent post was added.

Now over to the story itself. Set against the black background, the story should more or less appear the same size as a typical smartphone. (Stories are designed to be created primarily from mobile phones.) Don't worry if you don't have a smartphone; we go over how to create stories from your computer later in this chapter. Click the next icon (right arrow) to see the next post in the story. When the story you're looking at runs out of posts, you can proceed to the next story until there are no more. Click the back icon (left arrow) to return to a previous post or story. When the story ends, clicking the next icon displays the next person's story. To close Story Viewer and return to your normal News Feed, click X or the f logo in the top-left corner.

Facebook treats stories like slideshows, which automatically play or progress as soon as you open them. The thin grey bar at the top of the story tells you how many posts are in a story and where you are in the story. This bar is broken into segments to reflect the number of posts in the story. These segments fill with white as the story progresses, just like the progress bar in a video. If the story is progressing too quickly for you to keep up, click the pause icon below the grey bar. You can also turn the volume on and off or click the three dots icon to report the story.

Below the story is a bar for comments and reactions. Comments and reactions work pretty much the same in News Feed and stories, and are covered in the next section.

Interacting with Your News Feed

Unlike the newspaper on your doorstep in the morning, News Feed is not just a method of delivering news. It's more of a starting place, meant to facilitate more interactions between you and your friends. The bottom of each post has at least two options: Like and Comment. Many posts also have a third option: Share. Each of these options allows you to interact with your friends and their content. You can also save content to go back to later.

Liking

Liking is one of the simplest actions on Facebook. Here's how you like something:

1. Click Like.

Liking really is that easy. Anytime you see something you enjoy, you can click Like to let the person who shared it know that you liked seeing it. Like buttons may appear as text or as a thumbs up icon (or both). When you like something, the person who shared it will be notified and other people seeing that post may see that you've liked it. Being notified that someone liked something you shared is a great feeling. If you ever like something by accident, simply click Unlike to undo it.

Liking is meant to be simple, so we won't over-explain when you might like something, when you might like something instead of commenting on it, when you might like something *and* comment on it. Suffice it to say that if you enjoyed something you saw, or you agree with something your friend said, or, well, you liked the content, clicking Like communicates the sentiment loud and clear.

Reacting

Liking is the simplest way to let someone know you saw and appreciated something they posted. Sometimes, however, like isn't appropriate. People often post stressful or sad things on Facebook; it would be insensitive to respond to "My dog had to be put down today" with a like response. On the other side of things, sometimes a like isn't a strong enough term. Do you like that baby smiling, or do you *love it so much you want to eat its face?* Historically, you might see people commenting on these sorts of posts with one-word responses: Love. Dislike. LOL. Sad. A *reaction* is Facebook's way of providing more than just one button to express your sentiment. To leave a reaction other than a like, simply hover the mouse cursor over the Like button or link and wait for the Reactions list to open. You can then choose your Reaction from the following options:

 » **Like:** The old standby. Click the thumbs up icon to let your friends know you liked their post.

 » **Love:** The heart icon lets your friends know you loved their post.

 » **Care:** The hugging heart icon lets your friends know you're giving them a virtual hug.

 » **Haha:** The laughing smiley face lets your friends know you thought what they posted was funny.

 » **Wow:** The smiley face gaping in awe lets your friends know that you were impressed if not flabbergasted by their posts.

 » **Sad:** Express a little empathy by clicking the crying smiley face. It's the virtual equivalent of a gentle pat on the back.

 » **Angry:** The virtual equivalent of a little "grrrrr," clicking the glowering smiley face lets your friends know that you're angry.

When you click any of these reactions, your friend receives a notification that you reacted to his or her post.

Liking (or following) Pages

You can like almost anything on Facebook. You can like a photo or a status; you can even like a comment on a photo or status. But there's a slight difference between liking this sort of content and liking Pages.

Pages are official profiles that companies, bands, and public figures make to represent themselves on Facebook. They work mostly like timelines (the key differences are covered in Chapter 14), except instead of friending Pages, you like (or follow) them.

This sort of liking (or following) has one big implication. You may start seeing posts and updates from the Page in your News Feed, alongside posts from your friends. These sorts of updates can be interesting and cool if you're into the company or brand (for example, Disneyland or the *New York Times*). If they start to bother you, however, you can always hide that Page from your News Feed or unlike (or unfollow) the Page.

Commenting

Liking something is the quickest and easiest way to let your friend know that you saw what they had to say and enjoyed it. Commenting is also simple, and it takes you from a reaction — I liked this! — to a conversation. The only requirement for a comment is that you have to have something to say. The comment box appears under most content on Facebook. You can see an example of it in Figure 4-8.

Kara Lynne was 🍜 eating **noodles**.
September 25 · 👥   ...

The ramen was better in Japan

2 Comments

👍 Like   💬 Comment

Carolyn Abram
have you tried that new place up by the mall? it's really good!
Like · Reply · 1m

Kara Lynne
no...i'll have to check it out
Like · Reply · 1m

Write a reply...   😊 ☺ 📷 GIF 🎨

Write a comment...   😊 ☺ 📷 GIF 🎨

FIGURE 4-8:
Share your own thoughts with a comment, and it may spark a conversation.

REMEMBER

When you comment on something on Facebook, anyone who can see that item — whether it's a post, a photo, or something else — will also be able to see your comment and respond to it. By the same token, you can see comments from people you aren't friends with.

Adding a comment

Commenting isn't much harder than liking something. To comment on anything on Facebook, follow these steps:

1. Click Comment, if necessary, to expand the comment box.

2. Click in the text box that appears.

3. **Type what you want to say.**

4. **Press Enter.**

Your comment appears below the post. Whoever posted the item will be notified of your comment and will be able to respond.

After you comment on something, you'll be notified about subsequent comments so that you can keep up on the conversation.

Adding a comment with extras

Adding a comment is meant to be easy, but sometimes you need more than words to express yourself. For example, Carolyn once posted a photo of a fully unrolled tube of toilet paper, courtesy of her 2-year-old. The comments from her friends included their own photos of toddler toilet-paper-mayhem. You can add emojis, photos, *GIFs* (quick animations on loop), and stickers to comments as follows:

>> **Emojis:** Click the smiley face icon in the comment box to open a list of emojis to add to your comment. Emojis range from faces to flags to food. Simply click the emoji you want to add, and it will appear in the comment box.

>> **Photos:** Click the camera icon to open a window that allows you to select photos and videos from your computer's hard drive. Click the photo or video you want to add, and then select Choose or Open at the bottom of the window. When you post your comment, the photo or video will be included.

>> **GIFs:** A *GIF* is technically a type of file format for images, but people also use it to refer to short animated clips that play in a loop. These clips are often pulled from pop culture (a repeated loop of a character from a TV show rolling her eyes, for example) and can be used as visual shorthand or punctuation in text. Click the GIF icon in the comment box to open a list of GIFs you can add to a comment. These are sorted by trending, meaning the first few GIFs you see are the ones currently being used the most. You can use the search box to find a GIF for virtually any emotion, complex thought, or cultural meme. Simply click the GIF you want to post it as a comment.

>> **Stickers:** Like emojis and GIFS, stickers are a way for you to add an extra visual element to your comments. Stickers are meant to be the same as their real-world counterparts — a decorative little image that brightens someone's day. Click the sticker icon (a peeled sticky note) to open a list of stickers. You can pick a category (such as Happy or Sleepy) or search by keyword for the sticker you want. Click the sticker to post it as a comment.

One other commenting extra you can try is tagging a friend in your comment. *Tagging* is a way of creating a link between something you post and your friend's timeline. In this case, people often tag friends to bring their comments (or the original post) to the attention of other friends, because tagging someone sends the person a notification. You can tag a friend in a comment by typing the @ symbol (that's Shift+2) and then typing your friend's name. Facebook will autocomplete as you type, and you'll be able to select the person's name from the list that appears.

Editing and removing comments

If you decide your comment was a poor choice, all is not lost! You can edit or delete a comment at any time:

1. **Hover your mouse cursor over the comment you'd like to change or remove.**

   A small three dots icon appears to the right of the comment.

2. **Click the three dots icon.**

   A menu appears with two options: Edit and Delete.

3. **Select Edit to make changes to your original comment or Delete to simply remove it.**

   If you choose Edit, the comment box will reopen as if you had just finished typing your comment but not yet clicked Enter. If you choose to delete, you need to confirm that you want to delete the comment.

TIP

You can delete comments friends have made on your own content by following these same steps.

REMEMBER

If you make multiple edits to a comment, you can see all your previous edits by clicking the Edited link under that comment. Others may be able to see that same Edited link on your comment, especially the person who posted the content, but they won't be able to click or see any changes you've made.

Replying to and liking comments

As mentioned, commenting on something is a way of starting a conversation, and the way to continue that conversation is by liking or replying to comments. At the bottom of any comment are two small links to Like or Reply.

Liking a comment is generally a way of saying "I agree" or "Right on!" or "That's funny." Click Like next to any comment to let the commenter know how much you enjoyed what he or she had to say.

Replying to a comment is as easy as clicking the Reply link below the comment you want to respond to. This opens a text box that says Write a Reply . . . . Then follow the same commenting steps detailed previously.

Replies show up indented from the original comment, as shown previously in Figure 4-8.

Sharing

You've probably noticed the word *share* being used a lot on Facebook. In addition to the share box at the top of your News Feed and timeline, Facebook has a specific share feature, designed to make it easy to post and send content that you find both on Facebook and on the web.

TIP

To learn how to use the share box at the top of your News Feed, check out the "Sharing Your Own News" section of this chapter.

Perhaps you've already noticed the little Share links all over Facebook. They show up on albums, individual photos, events, groups, News Feed posts, and more. They help you share content quickly without having to copy and paste.

If you're looking at content on Facebook that you want to show someone, simply click the Share link near it. A menu appears, as shown in Figure 4-9, with several options:

» **Share Now (<*Privacy*>):** Choosing this option is the quickest way to share something and simply posts the content to your own timeline (and, by extension, your friends' News Feeds). The <*Privacy*> notice reminds you who normally sees content that you share on Facebook. It says Public or Friends or Custom, depending on your privacy settings.

» **Write Post:** This option opens a full share box that allows you to add your own commentary on the content you're sharing before you share it.

FIGURE 4-9: Share here.

- » **Send in Messenger:** This option opens a message box where you can type names you'd like to message with your friend's post. You can also add a comment to the Say Something about This section to let your friends know why you're sharing the content with them.

- » **Share to a Group:** This option allows you to post the content to a specific group you're a member of. It opens a share box similar to the normal share box, enabling you to add your commentary and select from a list of groups you belong to. You can find more about sharing with groups in Chapter 10.

- » **Share to a Page:** If you're a Page owner or admin, you can use this link to share something as a post from your Page. When you click this option, you can add commentary and decide if you want it attributed directly to the Page or you. Followers of your Page will then be able to see that content.

- » **Share on a Friend's Profile:** This option is the same as copying and pasting a link into a post you leave on your friend's profile (but it's much easier than all that copy/paste nonsense).

If you click Share on a friend's post, the friend who originally shared it is given a credit. So if you reshare an article, the post that your friends see will say Shared <*Friend's Name*>'s Post so that everyone knows where you found it.

If you're a reader of blogs, a viewer of videos, or an online shopper, you probably know that virtually everything you look at has links to share it on a variety of platforms, including Facebook. Next time you're reading an article on your favorite site, look to see if you can spot a Facebook logo (you may need to click Share first, and then select Facebook from a menu of options).

When you choose to share on Facebook, a share box opens in its own window. You can then view the post, add comments, adjust the privacy, and post it to your timeline without ever having to leave the page you were on.

Saving

News Feed can be one of the best sources for articles, videos, and legitimate old-fashioned news. If you follow several Pages for newspapers and magazines, and if your friends are the nerdy type who read and share articles that they find interesting as well, you might find that everything in your News Feed looks interesting and you don't have time to read it all.

That's why the Save feature is such a useful (if slightly hidden) Facebook feature. Saving articles, videos, and other posts is an effective way to make a reading list for later, when you have more time. It's also a good way to bookmark helpful websites you might need later (such as a list of the best pumpkin patches to visit in the fall).

To save a post, click the three dots icon in the upper-right corner of the post you want to save. The News Feed menu appears, as shown in Figure 4-10. Click the Save Post option (usually at the top of the menu) to add the post to your Saved section. You can get to your saved posts at any time by clicking Saved in the left sidebar. Facebook may also occasionally show you a selection of your saved items in your News Feed in case you forgot about them.

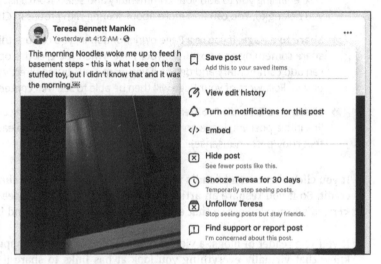

FIGURE 4-10:
The News Feed menu for a specific post.

Adjusting News Feed

News Feed is designed to learn about what you like and whom you care about and to show you posts accordingly. As you use News Feed, it learns what you like based on your clicks, comments, and likes. You can give News Feed more information to work with to make it even better for you.

Hiding posts and people

There are ways to tell News Feed explicitly who you do and don't want to see (and we go over these in the "News Feed preferences" section). But chances are you might not know what you like or don't like until you see it. Sometimes you'll realize that you'll simply explode if you have to see another baby photo from a particular coworker. Other times someone you know just won't stop posting political screeds. Or a post will just bug you and you won't want to see it anymore.

As soon as you realize that you don't like what you are seeing, you can easily give this feedback to News Feed by using the menu of options found as part of any News Feed post. To open this menu (refer to Figure 4-10), click the three dots icon in the upper-right corner of any News Feed post.

This menu gives you four options regarding the post you are viewing:

» **Hide Post:** Selecting this option immediately hides that post from your News Feed. Additionally, Facebook will try to show you fewer posts like this. In other words, depending on the type of post you're hiding (such as a photo post or life event) Facebook will try not to show you as many of that post type.

» **Snooze *<Friend or Page Name>* for 30 days:** Selecting this option immediately hides all posts from this friend or Page for the next month. Think of it as a trial run for unfollowing. After 30 days, you may see the friend's or Page's posts again in News Feed, at which point you can decide if you want to unfollow.

» **Unfollow *<Friend or Page Name>*:** Selecting this option semipermanently prevents posts from this friend from appearing in your News Feed. It's semipermanent because you may undo this option later if you want.

» **Find Support or Report Post:** If you think a post is offensive or might violate Facebook's terms, you can ask for help or report it directly from this menu. After you select this option, Facebook opens a Report dialog that lets you provide more information about why you're reporting (for example, the post contains hate speech), as well as options to make sure whomever posted it can be taken out of your News Feed, if you so choose.

Selecting any option will hide the post you're looking at and display a small confirmation message to let you know that you've successfully hidden the post or unfollowed the friend or Page. You can click the blue Undo here if you regret your actions.

If you're looking at a post that involves multiple people, such as a post about Moira (your friend) tagged in a photo by Roland (not your friend), you may see an additional option to Hide All from Roland. If someone is sharing a post from a Page or an app, you may also see options to hide posts from that Page or app.

REMEMBER

Unfollowing friends is different from unfriending them. *Unfriending* them severs the link between your timelines and may mean you are no longer able to see certain parts of their timeline. *Unfollowing* simply removes their posts from your News Feed.

News Feed preferences

Hiding posts and people is a good way to incrementally adjust your News Feed over time. But if you're looking to see a bigger effect right away, you may want to adjust your News Feed preferences instead. Your News Feed preferences allow you to choose friends you want to see first, unfollow friends you already know you don't want in your News Feed, and reconnect with people you may have unfollowed in the past.

To get to the News Feed Preferences window, open the Account menu (the down arrow in the top bar). Then click Settings and Privacy, and choose News Feed Preferences from the menu that opens. The Preferences window shown in Figure 4-11 appears.

There are four sections of preferences you can use to influence News Feed.

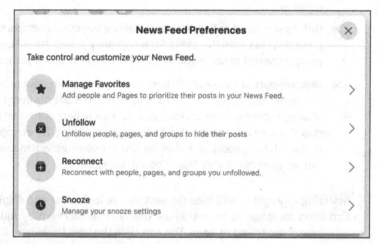

FIGURE 4-11:
The News Feed
Preferences
window.

Manage favorites

Adding friends to your favorites is a way of designating people you *always* want to see in News Feed. You may want to put your spouse, child, or best friend here. If you have a more distant friend whose posts you still love, adding that person to your favorites means you won't miss a thing.

Posts from prioritized friends will always be at the top of your News Feed when you log in. Unfortunately, if your friends don't post very often, there won't be anything to show you. News Feed needs something to work with. The only thing that differentiates such a post (other than the fact that it's at the top of your News Feed when you log in) is a blue star in the upper-right corner.

Click Manage Favorites to open a window for choosing people from your friends list. Facebook displays the people it thinks you'll likely want to see first at the top; you may have to scroll down to find whom you're looking for. Click the star icon next to any friend's name and face to select that person. (Click again to deselect the friend.) When you're finished, click the back arrow in the upper-left corner.

Unfollow

By default, when you become friends with someone you also start following them. This means their posts may appear in your News Feed. However, not all friendships are created equal, so unfollowing is a simple way to stop seeing posts while still maintaining the friendship. You don't need a particular reason to unfollow people (or unfollow a Page). You can do it because they post too frequently, you don't like what they post, or they just broke your best friend's heart.

To select people (and Pages) you want to unfollow, click Unfollow in the News Feed Preferences window. A list appears of all the people, Pages, and groups you currently follow. Click the blue check mark next to a name to unfollow. Click the back arrow when you're finished.

Reconnect

If your News Feed seems stale or bland, consider refollowing people you've unfollowed. Clicking Reconnect in the Preferences window opens a list of people whom you've unfollowed. Click the little folder with a plus sign to add them back into the News Feed mix. Click the back arrow when you're finished.

Snooze

Click Snooze to view all the people, Pages, and groups you've snoozed. (Remember, *snoozing* means you won't see posts from that person for 30 days.) If you find yourself missing their friendly faces, click the clock icon next to each name to end the snooze early.

Sharing Your Own News

While there are millions of things to do on Facebook (and this book tries to cover every single one of them), the most basic action on Facebook is sharing. The previous parts of this chapter cover how you see and interact with the things your friends have shared. Now it's time to put yourself out there and start sharing things yourself.

First things first: Take a look at *share box* at the top of your News Feed and shown in Figure 4-12. A similar box appears at the top of your timeline and most groups and Pages. You use the share box to create and share posts with your friends. Any posts you make from the share box will go on your timeline and may appear in your friends' News Feeds.

FIGURE 4-12:
Start sharing
here.

Like much of Facebook, a lot is packed into this little box. Before going into all the details of what you can share and how, we start with the most basic type of post: a status update.

Status updates

A *status update* is a way of reporting what's going on with you right now. Facebook uses the phrase "What's on your mind?" to prompt a status update. So you can share what you're thinking about, what you're doing, or really anything. To update your status, follow these steps:

1. **Click the share box, where it says, "What's on Your Mind, <*Your Name*>?"**

   The share box expands and the rest of the screen fades away so you can focus on your post, as shown in Figure 4-13.

2. **Type your status.**

   If you're experiencing writer's block, check out the next section, "Figuring out what to say."

3. **Click Post.**

   The Post button is at the bottom of the share box and will turn blue as soon as you begin typing.

Like we said, sharing on Facebook is easy.

After you've shared your status, your friends will likely see it in their News Feeds and will be able to like and comment your status.

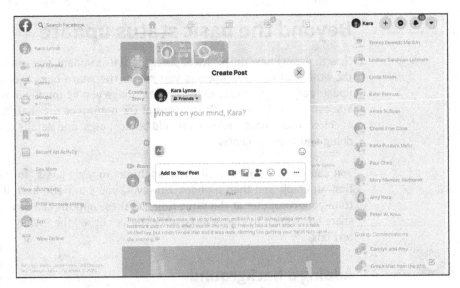

FIGURE 4-13:
The share box,
expanded.

Figuring out what to say

There isn't a simple answer to the question of what you should or shouldn't post as a status update. The question "What's on your mind?" is meant to provoke whatever it is you'd like to share.

In general, sharing where you are or what you're doing makes sense as a status update. When Carolyn sees updates from her friends, they're often a random thought that passed through their heads as they were out and about during the day. If a football game is on, she can often figure out what's happening based on her friends' status updates. Check out your own friends' posts to see what they're talking about.

You might share something notable you saw recently or tell a quick and funny story (or a longer one, as there isn't a word limit on status updates) about your day. Originally, status updates all started with the words "<Your Name> is. . ." and for the most part status updates tend to be about what you're doing or thinking right now.

All the examples captured in the figures in this chapter are the sorts of posts we see regularly on Facebook. If you're feeling uninspired, feel free to post that you're reading *Facebook For Dummies*.

Beyond the basic status update

Now that you know the basics of posting a status update, you can learn more about the other buttons and options you see in the share box. Adding links to your updates allows you to share interesting things you've found all over the Internet. You can also click any of the buttons at the bottom of the share box to add a pop of color to your post, and you can add photos, tags, and details about what you're doing and where you are.

You can combine almost all these additions to your posts, for example, adding first an emoji and then a photo. However, you can't add both links and photos from your computer at the same time. All these options can be accessed from the share box shown in Figure 4-13, after you've clicked the "What's on your mind?" text box.

Adding a background

Adding a background color or design is a fun way to add personality to a post. When you add a background color, it also changes the format and font size of your post, emphasizing your words. You can see the difference a background makes to a status update in Figure 4-14.

To add a background, follow these steps after you've typed your status:

FIGURE 4-14:
Adding a background gives your words a little more pop.

 1. **Click the colorful Aa square in the lower-left corner of the share box, below the text you've typed.**

Background previews appear below your text as a row of squares.

2. **Choose a background as follows:**

- *If you see a background you want:* Click your background choice.

- *To choose from more options:* Click the icon with a grid of four squares. In the Choose Background window that appears, scroll and select a background.

3. **Finish typing your post.**

   You can still edit your words or repeat Steps 2 and 3 to choose a different background.

4. **When everything is to your liking, click Post.**

Adding links

To add content from another website to your post, simply copy and paste the link into the share box. Facebook will automatically generate a preview for the content based on the link. Previews usually include an image, a headline, and a description. Depending on the content you're sharing, you may be able to choose from different thumbnail photos. You can also hover your cursor over the thumbnail image and click the X that appears in the upper-right corner to remove the article preview and share only the link.

Once you get your preview looking the way you want, you can delete the original link (the preview will remain) and use the space above to share your thoughts about your link. When you're ready to share, click Post.

Adding an emoji

An *emoji* is a smiley face icon or other small-sized image that you can insert into your post. To select from a list of emoji, click the grey smiley face icon on the right side of the share box. (The yellow smiley face icon is used to add feeling or activity words to your post.) When you see the emoji you want to add, click it. You can add emoji to text or simply create a post made entirely of emojis, as shown in Figure 4-15.

Kara Lynne
Just now ·

👍 Like   💬 Comment   ↪ Share

Write a comment...

FIGURE 4-15:
Emojis help
you express
everything
you feel.

Adding a photo or video

Click the green photo/video icon to add a photo to your post. When you click this icon, it opens an interface for navigating your computer's hard drive to find the photo you want to add.

Often some of the best photos you want to share on Facebook are on your phone. You can learn how to share photos directly from your smartphone in Chapter 7.

After you select a photo (or photos) you want to add, click Choose or Open. You return to the share box, and you see the photos as they'll appear in your friends' News Feeds. Hovering the mouse cursor over the photos displays three new buttons. Click the X in the upper-right corner to remove the photos from your post. Select the Edit or Edit All button to crop, rotate, caption, and tag individual photos. Click the Add Photos/Videos button to reopen the interface for selecting photos from your computer. For more information on these options, see Chapter 11.

When you're happy with the photos you've chosen, you can add more text or other information to the post, or simply click Post to share.

Adding a tag

A *tag* is a way of linking someone or something else to your content on Facebook. Most often, tags are used to let people know who is in a photo, but tags can be used also in status updates to let people know who is with you. People also use tags to bring certain friends' attention to something they're posting. For example, you might see a status update that says "Who's up for going hiking? **Patrick, Stevie?**" Each bolded name links to a friend's timeline. Additionally, those friends are notified when the post is published.

You can tag someone in your post in two ways. The first is to type the @ symbol (Shift + 2) and begin typing the name of the person you want to tag. Facebook autocompletes as you type. When you see your friend's name highlighted, press or tap Enter. When you tag someone this way, the tag appears as part of the post, such as "Way better than **Moira** at singing."

The second way that you can add a tag is by clicking the blue tag friends icon at the bottom of the share box. The Tag Friends window appears. Select your friend from the list that appears or type the person's name in the search box and press Enter when you see it highlighted in the autocomplete list. When you tag people this way, the name is appended to the top of the post: "Get practicing Jazzigals — with **Moira**." You can see both types of tags in Figure 4-16.

In addition to people, you can tag Pages. For example, you might want people to know that you're excited about the latest episode of *Schitt's Creek*. Type the @ sign and start typing **Schitt's Creek**, and you'll find that it appears in the autocomplete list.

Tagged using @ symbol   Tagged using tag icon

Kara Lynne is with **Amy Karasavas**.
September 25 ·

Amy Karasavas knows her floral notes

👍 Like   💬 Comment   ↪ Share

Write a comment...   😄 🙂 📷 GIF 🎁

FIGURE 4-16:
Tag your friends
when they are
with you.

Adding what you're doing or feeling

Clicking the yellow feeling/activity icon opens the How Are You Feeling? window on top of the share box. Here you can search and browse for choices to explain what you are doing, thinking, or feeling while writing your status update. You may be feeling blessed, or baking cookies, or traveling to the Grand Canyon. The options here are virtually endless.

The information you enter is appended to your post, often with an emoji that further illustrates what you're doing or feeling. Figure 4-17 shows an example of a post with information about what Kara's drinking, with coffee cup emoji to complement her drink of choice. Additionally, if something you're doing has a page or information about it on Facebook, Facebook may add a preview of that information to your post.

Kara Lynne is ☕ drinking **coffee**.
Just now ·

Wishing it was cold brew season...

👍 Like   💬 Comment   ↪ Share

Write a comment...   😄 🙂 📷 GIF 🎁

FIGURE 4-17:
A post with
activity
information.

You can add only one thing you're doing or feeling to a post, so unfortunately you can't be both feeling joyful and watching *Schitt's Creek* at the same time, as far as your posts go.

Checking in with your location information

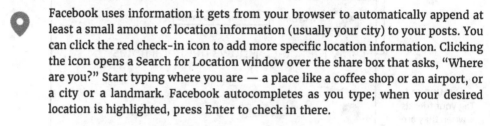

Facebook uses information it gets from your browser to automatically append at least a small amount of location information (usually your city) to your posts. You can click the red check-in icon to add more specific location information. Clicking the icon opens a Search for Location window over the share box that asks, "Where are you?" Start typing where you are — a place like a coffee shop or an airport, or a city or a landmark. Facebook autocompletes as you type; when your desired location is highlighted, press Enter to check in there.

Showing, not telling, with GIFs

Sometimes what you want to express is beyond mere words, or emojis, or photos from your camera roll. For those times, you can select a GIF, which is short video clip, often culled from television or pop culture, and post that to Facebook to let everyone know how you are truly feeling. Selecting the GIF icon opens the Choose a GIF window, which lets you search through the Facebook library of GIFs.

More options

Hidden behind the three dots icon are a few more extras you can take advantage of when making a post. When you click the icon, the Add to Your Post window appears over the share box, where you can choose from any of the previous or following icons:

>> **Live video:** Use the live video icon to broadcast live from your computer. You need a webcam to start a live video. You can learn more about live video in Chapter 11.

>> **Watch party:** Select the watch party icon to create a video playlist that your friends can watch in real time with you. It's a virtual way to have friends over to watch the season finale of your favorite show. When you create a watch party, you select the video(s) you want to watch from Facebook's collection of videos and then begin the party, which appears in your friends' feeds as a post. They will be able to react and comment as you all watch the video together.

>> **Raise money:** Select the coin icon to create a fundraiser to support a cause of your choosing, and use posts to your friends to promote the fundraiser. To discover more about fundraising for causes, see Chapter 12.

Creating a story

Stories are temporary posts that expire after 24 hours. Friends who log in within that time can see them, but friends who don't, won't. Stories are often meant to be added to over the course of a day, as bite-sized nuggets that dole out the, well, story of your day. Most people find it easier to add content throughout the day using their smartphones rather than their computers. Because of this preference, we describe stories in more detail when we cover using Facebook on a mobile phone, in Chapter 7.

However, you can create stories from your computer too. Just follow these steps:

1. **On your News Feed, click Create a Story.**

   This prompt appears at the top of the page, where your friends' stories are displayed. A page opens so you can create a photo or text story, as shown in Figure 4-18.

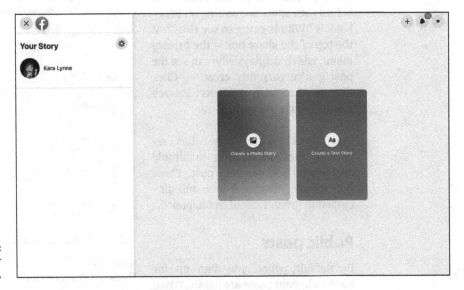

FIGURE 4-18:
Start telling your story.

2. **Choose Create a Photo Story or Create a Text Story.**

   For this example, we create a photo story, because photo stories are more common than text stories. An interface for selecting photos from your computer's hard drive appears.

3. **Select the photo you want to add and then click the Open button.**

   The photo loads in a preview page on the right.

4. **Edit the photo as needed.**

   You might want to crop the photo, rotate it, or add a text overlay (by clicking Add Text on the left). When you add stories from your phone, additional options appear.

5. **Click the blue Share to Story button of the left.**

   You return to your News Feed. The story preview appears at the top of the page, next to your friends' stories.

6. **To add more posts to your story, repeat Steps 1-5.**

Stories are meant to be updated frequently, often telling the story of your day. If your first update is about the spider building a web outside your window, by all means add to the story if a bee gets tangled up in it!

Controlling who sees your posts

The big question people often have before they share something on Facebook is "Who is going to see this?" At the top of the share box is the privacy menu, which displays who can see the post you're currently creating. Click this menu to open the Select Privacy menu, shown in Figure 4-19.

The Select Privacy menu has six options that allows you to simply decide who will see your post. These options are introduced here and discussed in greater detail in Chapter 6.

Public posts

By default, when you sign up for Facebook, your posts are public. When something is public, anyone can see

FIGURE 4-19:
The Select Privacy menu.

it. However, that doesn't mean that everyone will see it. Due to the volume of content on Facebook (two billion people adding even one post a day is a *lot* of posts) the chances of a true stranger seeing your post is low. Rather, the people in the public who will likely see your post are friends of your friends or other people whom you in some way interact with on Facebook — through a shared group, for example. If someone were to search you out and visit your timeline, that person would be able to see any public posts you've made there.

Friends-only posts

The second main option presented in the privacy menu is Friends. Posts visible to friends are visible only to people you've added as friends. You already know that when you post something, it may appear in your friends' News Feeds; the same holds true when you set your privacy to Friends. Friends of friends will not be able to see your posts, even if your friends comment on them. Someone searching you out and viewing your timeline would not be able to see your post. This privacy option is why we recommend that you never accept friend requests from strangers. When you know everyone whom you are friends with, it's easy to know who can see your posts.

Friends except posts

Even though we recommend that you accept friend requests from only people you would willingly share all your posts, sometimes that's just not possible. We all know what it's like to get a friend request from someone whom it's not acceptable to reject (love you, Mom!). Sometimes our posts aren't meant for their eyes. Choosing this option opens a window where you can type the names of friends whom you *don't* want to see your post.

Only me posts

Posts that have this setting are visible to, you guessed it, only you. Frankly, we don't think we've ever used this setting. If we don't want *anyone* to see something, we simply don't put it on Facebook.

Specific friends posts

If you want to share with a smaller group of people, you can choose individual friends to share with. This option might be helpful if you have a large friends list but want to share with only a fairly small group of people.

Custom posts

Custom privacy allows you to combine the Specific Friends and Friends Except options. From one window, you can decide who gets to see something and then exclude others.

Privacy is a way of controlling who can see what, but the easiest way to measure it is not in counting the number of people who can see something. If you have a huge friends list (and many people do) does the difference between something being available to 500 people or 1,000 people or 1,000,000,000 people matter?

Well, maybe. The way to measure privacy is not entirely numerical but also emotional. Do you feel comfortable? Do you feel safe? These are the questions you should answer when you're choosing between privacy settings.

REMEMBER

Whichever option you choose, the next time you go to post something, that same option will be selected. Facebook assumes that most of the time you're interested in sharing with the same people — either friends or everyone — so it doesn't force you to change your privacy each time you post.

Chapter 5

Timeline: The Story of You

One of the fun things about writing *Facebook For Dummies* is the process of going back over text that was written in the past and updating it based on whatever has happened in our lives. When Carolyn wrote the first edition, she was just out of college and almost every example related to Ultimate Frisbee, whether she was posting photos (of a Frisbee game) or planning an event (like a Frisbee match). By the second edition, she'd been working a little bit longer and used examples related to some fun travels. The third edition was all about her wedding, the fourth about her time in grad school, the fifth about her new baby, the sixth about her second baby, the seventh about her two toddlers, and for this edition, well, it turns out she's into her indoor plants (in addition to her children) right now. Your life can change a lot over the course of several years.

That moment of looking backward and seeing how far you've come is the idea behind the Facebook timeline. Like many websites, Facebook wants you to establish a profile with basic biographical information — where you're from, what you do, where you went to school. But Facebook also wants you to keep updating and posting and sharing and marking events that define you. Then it turns all that information into a virtual scrapbook that you and your friends can explore. That virtual scrapbook is your *timeline*.

On Facebook, your timeline and your profile are the same. We use the terms interchangeably. Your timeline shares a lot of traits with profiles you create on other sites: biographical info, links to your photos, friends, and so on.

This chapter covers all the ways you edit the information and appearance of your timeline, as well as who can see what on your timeline.

Making a First Impression

This section covers the very first things people see when they arrive on your timeline: your cover photo and your profile picture, both of which appear in Figure 5-1.

FIGURE 5-1: Welcome to your timeline.

These two photos at the top of your timeline present are your visitors' first impression of you. The *cover photo* is the larger photo that serves as a background to your timeline. Your *profile picture* is the smaller picture in the center of the page. This smaller photo, usually a headshot, follows you around Facebook and will appear whenever you make a comment or post.

WARNING

For this section, we assume that you added a cover photo and profile picture while setting up your account, so you're looking to change something that already exists. If you don't have a cover or profile picture yet, these same steps should more or less work, but the button may say Add Photo instead of Edit Photo.

Changing your cover photo

Because the cover photo takes up so much space on the screen, people often choose visually striking photos or images that speak to who they are and what they love. To change your cover photo, follow these steps:

1. **Click the Edit Cover Photo button in the lower-right corner of the cover photo.**

   The Update Cover Photo menu appears with four options: Select Photo, Upload Photo, Reposition, and Remove.

   If you're using the Facebook app on your smartphone, you may also see an option to Take a New 360 Photo. Choosing this option lets you take a panoramic view of your surroundings with your phone's camera.

   The cover photo spans the width of your timeline, so while uploading your photo, you might get an error stating that the photo isn't wide enough. Make sure your cover photo is at least 720 pixels wide to ensure that it will fit. Check the photo's info or dimensions (you may need to right-click or Control-click the photo first). As a general rule, vertical photos taken on your phone won't be wide enough.

2. **Select a cover photo from photos in Facebook or on your computer:**

   - *To select a photo from photos you've added to Facebook:* Click Select Photo. The window in Figure 5-2 appears, displaying Recent Photos by default. To see all your photos, click Photo Albums. Select your cover photo by clicking it.

   - *To select a photo from your computer:* Choose Upload Photo. A window for navigating your computer's files appears. Select your cover photo and then click Open.

   As soon as you select your photo, you return to your timeline, where you should see the new cover photo in place with the overlaid message, Drag to Reposition Cover.

3. **Click and drag your cover photo to position it correctly within the frame of the screen.**

4. **Click Save Changes.**

   Your new cover photo is now in place.

If you don't like the way your cover photo is positioned, click Edit Cover Photo to reopen the menu. You can then choose to reposition or remove your cover photo. You can change your cover as often as you want.

Select Photo

Recent Photos   Photo Albums

FIGURE 5-2:
Choose a cover
photo.

Editing your profile picture

Your profile picture is the smaller photo that greets your friends when they come to your timeline. This photo is what sticks with you all around Facebook, appearing wherever you comment or post something. For example, your friends may see your status post in their News Feeds, accompanied by your name and profile picture.

Most people use some variation of a headshot for their profile picture, though of course plenty of people feel they are better represented by cartoons, a headshot of their baby, or images other than their face.

When you first signed up for Facebook, you were likely prompted to add a profile picture. Whether you did or not, you can follow these steps to add a new profile picture at any time:

1. **Click the camera icon that appears in the bottom of your profile picture on your timeline.**

The Update Profile Picture window appears, as shown in Figure 5-3. Facebook sorts photos into categories: Suggested Photos, Uploads, and Photos of You. You can click the See More button below any of these categories to look at more photos. The main emphasis in this window is photos on Facebook, but you can also use the Upload Photo button or the Add Frame button to choose a photo from your hard drive or add a frame to your existing photo, respectively.

Update Profile Picture

+ Upload Photo   □ Add Frame   ✏

Suggested Photos

See More

Uploads

See More

Photos of You

FIGURE 5-3:
Choose a new
profile picture
here.

FIGURE 5-3:
Choose a new
profile picture
here.

TIP

If there aren't any photos of you in the Update Profile Picture window, it's likely because tagged photos of you haven't been added to Facebook yet. You'll still be able to add a photo from your computer's hard drive.

2. **Select a photo:**

   • *To select a photo on Facebook:* Simply click the photo in the window.

   • *To select a photo on your hard drive:* Click the Upload Photo button, at the top of the screen. In the window that appears, navigate to and select the photo.

   Options for editing the picture appear.

3. **Adjust the zoom and position your profile picture as needed.**

   You can zoom in or out on the photo as well as move it around to make sure your face fits nicely in the circular frame. You can also add a description, which is like a caption for this particular photo.

TIP

When you're editing your new profile picture, look below it for a Make Temporary button. People often want to change their profile picture just for a day or two. For example, you might want to celebrate a wedding anniversary by changing your profile picture to a wedding portrait for a day. After you click the Make Temporary button, you choose the date to switch back to your regular profile picture.

4. **Click Save.**

   You return to your timeline, where your new profile picture is waiting for you.

Adding a frame to your profile picture

Often people use their profile pictures to show solidarity with something that's happening in the world, from expressing sympathy for victims of a natural disaster to supporting certain legislation or rooting for your alma mater during football season. For this situation, frames are the easiest way to modify your profile picture.

Frames are just what they sound like — something that goes around your picture. Frames are created by outside developers, submitted to Facebook, and assuming they adhere to Facebook's guidelines and policies, available for you to add to your profile picture.

To add a frame to your profile picture, follow these steps:

1. **Click the camera icon that appears in the bottom of your profile picture on your timeline.**

   The Update Profile Picture window appears (refer to Figure 5-3). Across the top are three options: Upload Photo, Add Frame, and Edit Thumbnail (pencil icon). The bottom part of the window displays photos on Facebook that you may want to choose as your profile picture.

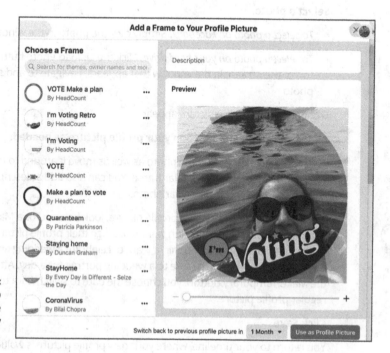

FIGURE 5-4: Add a frame to make your profile picture say something.

2. **Click the Add Frame button.**

    An interface for choosing frames appears, as shown in Figure 5-4. Facebook displays the most popular frames by default on the left side of the screen.

3. **Click a frame to see what it will look like on your profile picture.**

    You can use the search box at the top of the interface to look for something specific, such as a particular sports team. When you choose a frame, a preview appears on the right.

4. **Reposition the picture in the frame as desired.**

    You can click and drag your profile picture to reposition it within the frame, as well as use the slider to zoom in and out.

5. **When you're happy with your frame, click Use as Profile Picture.**

    By default, frames expire after one month, after which your picture returns to its frameless state. You can change how long your frame will stick around by using the drop-down menu next to the blue Use as Profile Picture button.

Much like your cover photo, you can change your profile picture as often as you choose. Every photo you select as your profile picture is automatically added to the Profile Pictures album.

REMEMBER

Your cover photo and profile picture are visible to anyone who searches for you and clicks on your name. Make sure you're comfortable with everyone seeing these images.

Adding a bio

The last part of a visitor's first impression of your timeline is the small space below your name, where you can add a short (101 characters) bio. The bio can be anything you want: a literal bio like you'd put on a resume, a favorite quote, a joke, a list of favorite movies, whatever it is you want people to know about you right away.

To add a bio, click the Add Bio link (or if you already have a bio, click the Edit link below it it). In the box that appears, type what you want to share, and then click the Save button.

Telling Your Story

Getting back to the main focus of your timeline, take a look at the what appears below the cover photo. Two columns run down the page:

» The skinny column on the left side contains some biographical info, as well as sections about your friends and photos. We describe elements in this column in the "Checking out your intro and more" section, later in this chapter.

» The wider right column is where posts, including life events, live. The posts might be something you've added to Facebook, such as a status or a photo, or something someone has added to Facebook about you, such as a photo tag. These posts constitute your timeline. As you scroll down, you can see what you were posting last week, last month, and last year.

In terms of navigating the timeline, the most important thing to know is that you scroll down the page to go back in time. As you scroll down, posts you and your friends have made and life events you have added keep showing up. This section goes over the basics of sharing your story, from the ongoing process of status updates and photo posts to posting life events to curating your timeline to highlight your favorite posts and events.

Creating posts

Posts are the type of sharing you'll be doing most often on Facebook. To post content — statuses, photos, places, links, and so on — to your timeline, you use the share box. The share box is the text field at the top of your timeline's right column, as shown in Figure 5-5. When you post content, you can also choose who can see it. Friends and followers (if you allow them) may then see these posts in their News Feeds when they log in.

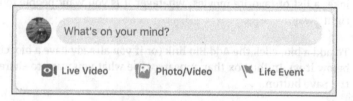

FIGURE 5-5:
The share box on the timeline.

What's on your mind?

Live Video   Photo/Video   Life Event

The share box on your timeline is similar to the share box at the top of your News Feed. Both are used to create posts. You can use either one to add a status, photo, video, or link as well as tags, emotions, and location information. The main difference is that the share box on your timeline enables you to create life events and to change the date of your post to add content (such as photo albums) to the proper point in your history.

The most common type of post is a basic text update that answers the question, "What's on your mind?" Posts like this are quick, short, and open to interpretation. People may say what they're doing at that moment ("Eating a snack"), offer a random observation ("A cat in my backyard just caught a snake!"), or request info ("Planning a trip to Chicago this summer. Anyone know where I should stay?"). It's easy for friends to comment on posts, so a provocative update can get the conversation going. We comment on commenting in Chapter 4.

Individually, these posts may sound small and inconsequential, but altogether they can tell a big story. For close friends, statuses keep you up-to-date on their daily lives and let you share a casual laugh over something you might never hear about otherwise. As a collective, posts are how news spreads quickly through Facebook. Because your posts go into your friends' News Feeds, a single update can have a big effect. For example, news of a minor earthquake in a local area spread faster on Facebook than it did on news sites.

Basic instructions for creating a post follow. For a more detailed dive, check out Chapter 4:

1. **Click in the What's on Your Mind field of the share box.**

   The Create Post window opens.

2. **Type your comment, thought, or status.**

3. **(Optional) Add any of the following details to your post:**

   - *Background color:* If you want your post to appear on a colorful background instead of the usual black-on-white text box, click the multicolored square in the lower-left corner of the Create Post window.

   - *Emojis:* If you want to add emojis to your post, click this icon to browse Facebook's selection and find the one that best fits your post.

   - *Photo or video:* You can add photos or videos from your computer's hard drive by clicking the green picture icon and selecting the photos you want.

   - *Tags:* Use a tag to mark people you're with when you're writing a status update. The tag links back to your friend's timeline and notifies him or her of your update.

     If you want to tag a person as part of a sentence as opposed to just noting that he is with you, add the @ symbol and begin typing the person's name. Facebook autocompletes as you type, and the tag appears as part of your status update: for example, <Eric> kicked my butt at Settlers of Catan.

TIP

- *Doing, thinking, feeling information:* You can add details to your status about what you're reading, watching, listening to, feeling, doing, and so on. Click this icon, select from the menu that appears, and append it to your post.

- *Location info or check-in:* If you click the pin icon and begin typing a city or place name, Facebook tries to autocomplete the place where you are. A post with location info is often referred to as a *check-in.*

4. **(Optional) Click the privacy menu (just below your name) to change who can see this post.**

   You can choose from the usual options: Public, Friends, Friends Except, Specific Friends, Only Me, or Custom. Whatever you select is saved for your next status post. In other words, if you post a link to Friends, the next time you post something, Facebook assumes you want to share it with Friends. We go over post privacy, including more advanced privacy options, in Chapter 6.

5. **Click Post.**

If those steps made you feel like updating your status requires way too much work, please remember how many of those steps are optional. You can follow the abridged version of the preceding if you prefer:

1. **Click in the share box.**
2. **Type your update.**
3. **Click Post.**

Frequently, people want to bring attention to something else on the Internet, such as an article they found interesting, an event, or a photo album. Usually, people add a comment to explain the link; other times, they use the link itself as their status, almost as though they're saying, "What I'm thinking about right now is this link."

When you add a link to a post, you can share something you like with a lot of friends without having to create an email list, call up someone to talk about it, or stand behind someone and say, "Read this." At the same time, you're almost more likely to get someone to strike up a conversation about your content because it's going out to more people, and you're reaching a greater number of people who may be interested in it.

To post a link, simply follow the preceding posting instructions but copy and paste the link you want into the field where you normally type your update. Facebook generates a preview of the link in your post: a separate box with a headline, a thumbnail photo, and teaser text. You can add your own thoughts about the link to the space above the preview.

After the preview has loaded, if you delete the full link text from the share box, it doesn't remove the link from your post. In fact, deleting the link can make your post look cleaner and leave more room for your own thoughts about the link.

Creating life events

Part of what's nice about Facebook is the way it lets you connect with friends over the small stuff: a nice sunset on your walk home, a funny observation in the park (Cats in strollers! Hilarious!). But Facebook is also awesome for letting you connect over the big stuff. Babies being born, houses being purchased, pets being adopted. Life events help you make note these milestones on your timeline.

Although it's not required, you may also feel an urge to fill out your history on your timeline. If you're new to Facebook, you may want to expand your timeline back past the day you joined. Life events are a good way to think about what you want to add in your history.

To add a life event, follow these steps:

1. **From your timeline, click Life Event, which appears at the bottom of the share box.**

Be sure to click Life Event, not the text field in the share box.

The Create Life Event window appears, displaying a menu of various types of life events. Here are the categories; when you click each one, you'll find many subcategories as well:

- Work
- Education
- Relationship
- Home & Living
- Family
- Travel
- Interests & Activities
- Health & Wellness
- Milestones & Achievements
- Remembrance
- Create Your Own

Milestones can be big or small; if you check out the subcategories, you'll see milestones ranging from getting your braces removed to learning a new hobby to having a baby. Feel free to make up your own. Lots of people use life events to represent small accomplishments in a humorous way. Hence, successfully canceling your cable service or getting the pile in the laundry room down to zero could also be listed as a life event.

2. **Select the event you want to create.**

   A pop-up window appears with text fields to fill out and space for photos to accompany the event, as shown in Figure 5-6.

3. **Choose who can see this event in your timeline by clicking your audience in the Sharing To field (at the top).**

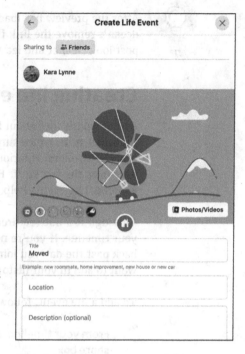

FIGURE 5-6:
Add a life event from the recent or distant past.

You have the same basic options as you have all over Facebook. The most common audiences are Public and Friends.

4. **Fill out the details you want to share.**

   In the example in Figure 5-6, which is a life event about moving, the details include the location being moved to and a description.

   You don't have to fill out all the fields, but some may be required to create the event.

5. **Add a photo to illustrate the event.**

   Facebook offers some preloaded illustrations that accompany your life event, but you can also add your own images from your computer's hard drive by clicking the Photos/Videos button.

6. **Adjust the date of the event if needed.**

   By default, events occur on the date you create them. If you're creating an event from the past, adjust the date by clicking the grey date button above the Share button.

7. **(Optional) Click the tag or location icons (next to the date) to tag friends or places to the event.**

8. **Click Share.**

   The event is then added to your timeline, and any photos you've added are featured prominently.

Editing posts

After you create a post, you may realize that you typed something wrong or want to add details. You can edit almost every part of a post you created after the fact. From your timeline, click the three dots icon in the upper-right corner of any post to reveal the Edit Post menu, shown in Figure 5-7. Keep in mind that depending on the type of post, you may see slightly different options in your Edit Post menu.

FIGURE 5-7:
Edit or
delete a post.

This menu displays slightly different options depending on the type of post, but for the most part, the Edit Post menu has at least the following options:

>> **Save Post:** Saving a post adds it to your Saved Items, which you get to from the left sidebar on your Home page. We usually use Saved Items to keep track of other people's posts that we want to check out at a later time, not for saving our own posts.

>> **Edit Post:** Choosing to edit a post reopens the share box so you can change any portion of the post, such as changing the wording of your status or removing a photo. Click Save to make your changes.

After you've edited a post, you can view the history of that post by clicking View Edit History from the Edit Post menu. Viewing the edit history lets you see all the changes you've made in the order you made them.

>> **Edit Audience:** Editing your audience means editing the privacy settings on the post, thus changing who on Facebook can see it. Choosing this option opens the Select Privacy window, where you can choose your preferred option for the post.

>> **Turn Off/On Notifications for This Post:** If you post something that a lot of people are responding to, you may find yourself inundated with notifications. You can turn off notifications for a given post if those notifications are annoying. You can turn them on again from the same Edit Post menu.

>> **Turn Off/On Translations:** If you have lots of friends and followers from all over the world, Facebook may attempt to automatically translate text in your post so everyone can read it. You can make sure it appears the way you wrote it by turning translations off.

>> **Edit Date:** Often you share items on Facebook after the date when they happened. For example, photos from a wedding might not be available to be shared until well after the wedding. Clicking Edit Data opens the Edit Date window, where you can change the date of a post so it shows up at the right spot in your timeline. Click Done to save your changes.

>> **Move to Archive:** When you archive a post, you keep that post from appearing in your timeline, but the post still exists in your archive, which is part of your activity log. The activity log which lists all your activity on Facebook, and only you can see it. When you move something to your archive, you essentially remove it from your timeline and hide it from your friends. But if you still want to see that post and its comments, navigate to your activity log (from the Settings and Privacy menu) and review the old posts in your archive.

>> **Delete Post:** Deleting a post gets rid of it from Facebook for good. Note that any photos in the post and any comments on the post are deleted as well.

If you're looking to remove photos or videos that exist only on Facebook, keep in mind that once they're gone from Facebook they're gone forever. It might be more practical to archive photos and videos instead of deleting them entirely.

Checking out your intro and more

The right column of your timeline features your posts and life events and updates every time you add something new. The left column of your timeline is more static and provides snapshots of some parts of your profile that people tend to look for when they visit a timeline:

>> **Intro:** The Intro box (see Figure 5-8) shows a portion of the information you may have added to your About section. The part it does show includes the things that help identify you as you.

TIP

Think of your intro as "dinner party introduction." These pieces of info — where you work, where you live, where you're from, who your spouse is — are the sorts of things you might talk about the first time you meet someone. To edit these parts of your Intro box, click the Edit Details button, or read the upcoming "Telling the World about Yourself" section.

The Intro box also includes featured photos and an option to add details about hobbies you enjoy. Featured photos are your most favorite photos. You can choose from photos you've added or that other people have added of you. To add a featured photo, follow these steps:

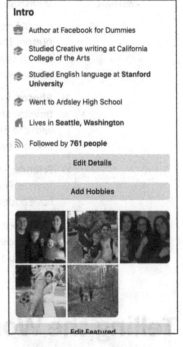

FIGURE 5-8:
Hi, nice to meet you; here are the basics.

1. **Click the Add Featured Photos (or Edit Featured Photos) button in the bottom of the Intro box.**

   The Edit Featured window appears, displaying thumbnails of photos of you on Facebook. You can also click the blue Upload Photo button to add photos from your computer's hard drive.

2. **Click your desired photo to select it.**

   The photo you've selected appears on the left side of the window. You can select up to nine featured photos.

3. **If you want to add more featured photos, repeat Step 2.**

   If you decide you don't want one of the photos you've chosen, hover your cursor over its thumbnail and click the X that appears in the upper-right corner.

4. **When you've chosen the photos you want, click Save.**

   Your photos appear in your Intro box as featured photos.

WARNING

Featured photos, like your cover photo and profile picture, are *public*, meaning anyone who visits Facebook, or who Googles your name and clicks your Facebook profile as a search result, can see them. Make sure they are photos you are comfortable with the world seeing.

If you'd also like to add hobbies to your Intro box, click the Add Hobbies button. This opens a window for selecting hobbies. Type your hobbies into the search box and click the matching hobby when you see it appear. Click Save when you're done.

>> **Photos:** In addition to featured photos, the Photos box on the left side of your timeline shows thumbnails of photos you've added or been tagged in, starting with the most recent. Clicking a photo thumbnail opens that photo in the photo viewer. You can then click to page through all your photos.

>> **Friends:** The Friends box shows thumbnail photos of, you guessed it, your friends. Friends with new posts will appear at the top, followed by friends you have added most recently.

>> **Life Events**: The Life Events box shows the most recently added life events you've posted on your timeline. You can click the See All link in the upper-right corner to see a full list of events that you've added to Facebook.

Telling the World about Yourself

The Intro box gives you (and your friends) the dinner party basics: where you live, what you do, where you're from, whom you're with. But Facebook gives you the opportunity to share a lot more information about yourself. Clicking the About link below your cover photo opens the expanded About section of your timeline, shown in Figure 5-9. By default, you see an overview of all your information — from here you can choose different sections to edit.

This page houses lots of information about you: work and education, places lived, contact and basic information, family and relationships, details about you, and life events. Much of this information won't change much over time, so it needs to be edited only when something big happens, such as a move to a new city. Click any of the sections to edit or add to the information there. You can also edit who can see this information, unlike the cover photo and profile picture. By default, your information is public, meaning everyone can see it. We go over changing your privacy in more detail in Chapter 6.

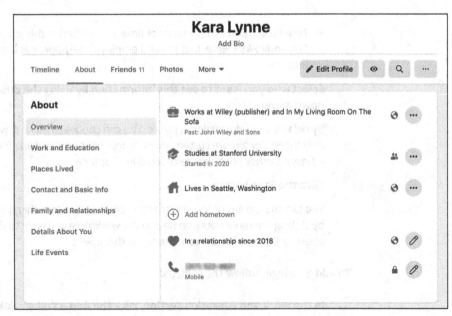

FIGURE 5-9:
The About section
of your timeline.

Adding work and education info

Your work and education information is important information to add to Facebook. This information helps old friends find you for reunions, recommendations, or reminiscing. It also helps people identify you as you, as opposed to someone else with a similar name.

To add an employer, follow these steps:

1. **In the Work and Education section, click the Add a Workplace link.**

   A form appears for adding company info.

2. **Start typing the name of the company where you currently work or worked.**

   Facebook tries to find a match while you type. When that match is highlighted or you finish typing, press Enter.

3. **Enter details about your job in the fields that appear.**

   These fields are as follows:

   - *Position:* Enter your job title.

   - *City/Town:* Enter where you physically go (or went) to work.

   - *Description:* Provide a more detailed description of what it is you do (or did).

- *Time Period:* Enter the amount of time you worked at this job. If you select I Currently Work Here, that job and employer will appear at the top of your timeline.

4. **Select who you want to see this information by using the privacy drop-down menu.**

   By default, this information is public. You can choose to make it available to only friends, or a more custom set of people or your friend list. These more advanced privacy options are covered in Chapter 6.

5. **Click the Save button.**

   You can change any of the information about this job (or others) in the future by clicking the three dots icon next to the workplace you want to change and selecting Edit Workplace from the menu that opens.

To add a college, follow these steps:

1. **In the Work and Education section, click the Add a College link.**

   A form appears for adding info about your college.

2. **Start typing the name of the college you attended (or attend).**

   Facebook tries to find a match while you type. When that match is highlighted or you finish typing, press Enter.

3. **Enter details of your school in the fields that appear.**

   These details include the following:

   - *Time Period:* Click the blue text Add Year to show when you started and finished your degree.

   - *Graduated:* You can select the graduated box to note that you received a degree (or deselect it to denote you did not).

   - *Description:* Add details about your time at school that you think may be relevant.

   - *Concentrations:* List any majors or minors you had.

   - *Attended For:* Choose whether you attended as an undergraduate or a graduate student. If you select Graduate Student, you'll have additional space to enter the type of degree you received.

4. **Select who you want to see this information by using the privacy drop-down menu.**

   By default, this information is public. You can choose to make it available to only Friends or a more customized set of people. These more advanced privacy options are covered in Chapter 6.

5. **Click the Save button.**

You can change any of the information you just entered by clicking the three dots icon next to the college and selecting Edit School from the menu that opens.

To add a high school, follow these steps:

1. **From the Work and Education Section, click the Add a High School link.**

A form appears for adding information about your high school.

2. **Start typing the name of the high school you attended (or attend).**

Facebook tries to find a match while you type. When that match is highlighted or you finish typing, press Enter.

3. **Enter details of your school in the fields that appear.**

These details are as follows:

- *Time Period:* Click the blue text Add Year to show when you started and finished your degree.

- *Graduated:* Select this box if you graduated from this school.

- *Description:* Add details about your time at school that you think may be relevant.

4. **Select who you want to see this information by using the privacy drop-down menu.**

By default, this information is public. You can choose to make it available to only Friends or a custom set of people. More advanced privacy options are covered in Chapter 6.

5. **Click Save.**

You can change any this information at any time by clicking the three dots icon next to the school you entered and selecting Edit School from the menu that opens.

Adding the places you've lived

The Places Lived section enables you to add your hometown, current city, and any other place you've lived. If you haven't previously added your current city or hometown, click the Add Current City, Add Hometown, or Add City link to do so. If you need to edit or remove your current city, click the three dots icon next to that city, and then select Edit or Remove <City> from the menu that opens. If you're editing your city, just begin typing your city name; Facebook autocompletes as you type. Select your city when you see it appear. Remember to click Save when you've successfully added or removed a city.

Adding contact and basic info

The Contact and Basic Information section contains just what you'd think: the very basics about you and how to get in touch with you:

>> **Phone numbers:** You can add as many phone numbers as you have — home, mobile, and work.

>> **Address and neighborhood:** You can choose to add this information in case anyone needs to mail you a present.

>> **Email address(es):** You can add as many email addresses as you want, and choose who can see those addresses.

>> **Website and other accounts:** You can add information about your account handle for other websites or services such as Skype, Instagram, and Twitter. You can also add a link to any personal websites you have.

You can add a new type or piece of contact info by clicking the blue links to Add *<Contact Info>*. To edit any piece of information listed here, click the pencil icon to its right. You can also edit who can see something by clicking the privacy icon (a globe, two people, or a lock, depending on your existing privacy). Clicking the privacy icon opens the Select Privacy window, where you can select your desired privacy level. When you've finished editing your information or privacy, click the Save Changes button.

It may seem a little scary to add your contact information to the Internet. If you're not comfortable doing so, that's okay. Facebook itself is a great way for people to reach you, so don't feel that adding other ways for people to contact you is required.

That being said, it can be useful for your friends to be able to find your number or address if needed. Privacy options (which are discussed in Chapter 6) can help you feel more comfortable sharing some of this information.

Your basic information includes the following categories:

>> **Birth date and birth year:** You entered your birthday when you registered for Facebook. Here, you can tweak the date (in case you messed up) as well as decide what people can see about your birthday. Some people don't like sharing their age, their birthday, or both. If you're one of these people, use the privacy menu to select what you want to share.

Although you can change your birthday and year at will, Facebook's systems prevent you from changing your birthday too often and from shifting to under 18 after you've been listed as over 18. If you signed up with the wrong birthday by accident and can't change it to the correct date, contact Facebook's Help Team from the Help Center.

>> **Gender:** You entered your gender when you signed up for Facebook, and Facebook mirrors your selection here. Are you transgender or have a preference as to how people refer to your gender? You can edit your gender by selecting Custom from the drop-down menu and entering whatever word best describes you in the text box. You can also choose what pronoun should be used in News Feed stories and around Facebook.

>> **Interested in:** This field is primarily used by people to signal their sexual orientation. Some people feel that this section makes Facebook seem like a dating site, so if that doesn't sound like you, you don't have to fill it out.

>> **Languages:** Languages might seem a little less basic than, say, your city, but you can enter any languages you speak here.

>> **Religion:** You can choose to list your religious views and describe them.

>> **Political views:** You can also choose to list your political views and further explain them with a description.

By default, your basic information (with the exception of your birthday) is public. Click the privacy icon next to any piece of information to change who can see it.

Adding family and relationships

The Relationship section and the Family section provide space for you to list your romantic and family relationships. These relationships provide a way of linking your timeline to someone else's timeline, and therefore require confirmation. In other words, if you list yourself as married, your spouse needs to confirm that fact before it appears on both timelines.

You can add a relationship by following these steps:

1. **From the Family and Relationships section of the About section, click Add a Relationship Status.**

   An area for adding this information appears.

2. **Click the Status drop-down menu to reveal the different types of romantic relationships you can add.**

These options are Single, In a Relationship, Engaged, Married, In a Civil Union, In a Domestic Partnership, In an Open Relationship, It's Complicated (a Facebook classic), Separated, Divorced, and Widowed.

3. **Do one of the following:**

   - *Stop here.* If you don't want to link your profile to the person you're in a relationship with (or are listing yourself as single), skip to Step 5.

   - *Link to the person with whom you're in this relationship:* Type the person's name in the box that appears. Facebook autocompletes as you type. Press Enter when you see your beloved's name highlighted. After you've finished these steps, this person will receive a notification about being in a relationship with you.

4. **(Optional) Add your anniversary by using the drop-down menus that appear.**

   If you add your anniversary, your friends will see a small reminder on their Home pages on that date.

5. **Click Save.**

   If you need to change your relationship status again, hover your cursor over it and click the pencil icon to edit your status.

TIP

For many couples, the act of changing from Single to In a Relationship on Facebook is a major relationship milestone. There's even a term for it: Facebook official. You may overhear someone saying, "It's official, but is it Facebook official?" Feel free to impress your friends with this knowledge of Facebook customs.

You can add a family relationship by following these steps:

1. **From the Family and Relationships section of the About section, click Add a Family Member.**

   A box for adding this information appears.

2. **Click the text box for Family Member and start typing your family member's name.**

   Facebook autocompletes as you type. When you see the name appear, click to select it.

3. **Select the type of relationship by using the drop-down menu.**

   Facebook offers a variety of family relationships ranging from the nuclear to the extended.

4. **Click Save.**

   Facebook sends a notification to that person.

Adding details about yourself

The Details About You section is a bit of a catch–all section for details that don't fit anywhere else:

>> **About You:** This section is a free-form place for you to describe yourself, if you so choose.

>> **Name Pronunciation:** If people are always mispronouncing your name, you can add a phonetic spelling of it. This pronunciation will be available on your profile with an audio component that people can play to learn the proper way to say your name. This information is always public and can be found by clicking the About section of a timeline and navigating to the Details About <Name> section.

>> **Other Names:** You can add nicknames, maiden names, or any other names people might use to search for you. This information is always public. Click the Show at Top of Profile option to display this name at the top of your profile next to your real name. This feature is especially helpful for maiden names.

>> **Favorite Quotes:** If you have any quotations you live by, you can add them to this free-form field.

>> **Blood Donations:** You can sign up here to be a blood donor. Facebook will notify you about blood drives near you.

Adding life events

You can see a summary of all the life events on your timeline in the Life Events section. If you see big gaps in your history that you'd like to fill, click Add a Life Event. The interface described in the "Telling Your Story" section appears. Follow the directions in that section as many times as necessary to fill out your summary to your satisfaction.

Viewing Timeline Tabs

In addition to the Posts and About sections of your timeline, which is where you and your friends will spend most of your time, other tabs on your profile have a specific type of information. These tabs are below your profile picture and above your Intro box.

The Friends tab, for example, is just a list of all your friends. The Photos tab displays all the photos you have added and been tagged in. Click the More tab below your cover photo to see even more tabs you can visit, including the following:

>> **Story Archive:** View stories you've posted that have since expired from Facebook. (A *story* is a special type of Facebook post that expires, or disappears from Facebook, after 24 hours.) No one else can see these.

>> **Videos:** See videos you have recorded and added to Facebook.

>> **Likes:** Look at Pages you've liked, starting with the most recent. You can also click to view likes by category (such as Restaurants or Books).

>> **Check-ins:** View locations where you've checked in recently.

>> **Events:** See Facebook events you have RSVPed to in the past.

You can decide which sections are part of your timeline. Click the More tab below your cover photo and select Manage Sections at the bottom of the menu that opens. The Manage Sections window appears. You can select and deselect sections to decide which ones appear as part of your timeline.

To edit the privacy of an individual section, navigate to that section (for example, select Movies from the More menu to go to the Movies tab). Click the pencil icon in the upper-right corner of the section to open the following options:

>> **Hide Section:** Hide the section from appearing on your timeline.

>> **Activity Log:** Go to the Activity Log and see all activity related to that section that could appear in that particular tab. Activity Log gets a little more attention in Chapter 6.

>> **Edit Privacy:** Edit who can see this section on your timeline.

Your Friends and Your Timeline

Your timeline is what your friends look at to get a sense of your life, and it's also where they leave public messages for you. In this way, your friends' posts become part of your history (just like in real life). Think about all the things you learn about a friend the first time you meet his parents, or all the funny stories you hear when your friend's significant other recounts the story of how they met. These are the types of insights your friends may casually leave on your timeline, making all your friends know you a little better.

When friends visit your timeline, they see the version of the share box shown in Figure 5-10. This version enables them to post text or a photo to your timeline. Check out the posts on your friends' timelines. Chances are, you'll see a few "Hey, how are you? Let's catch up" messages; a few "That was an awesome trip/dinner/drink" messages; and maybe a few statements that make so little sense, you're sure they must be inside jokes.

FIGURE 5-10:
Leave a public message for a friend on his timeline.

If you're on a friend's timeline around his birthday, you're sure to see many "Happy Birthday" posts. There aren't many rules for using Facebook, but one tradition that has arisen over time is the "Happy Birthday" post. Because most people see notifications of their friends' birthdays on their Home pages, the quickest way to say "I'm thinking of you" on their special day is to write on their timeline.

TIP

Although the back and forth between friends is one of the delights of the timeline, some people find it a little hard to let go. If you're someone who doesn't like the idea of a friend being able to write something personal on your timeline, you can prevent friends from posting. You can also limit who can see posts that your friends leave. From the Settings page — which you can get to from the account icon in the top bar — go to the Timeline and Tagging section and modify the settings related to who can post on your timeline and who can see what others post on your timeline.

REMEMBER

The best way to get used to the timeline is to start using it. Write on your friends' timelines, post a status update or a link on your own, and see what sort of response you get. After all, that's what the timeline is all about — sharing with your friends.

Chapter **6**

Understanding Privacy and Safety

When people talk about privacy online — and on Facebook in particular — we like to remind them that there's a spectrum of privacy concerns. On one end of the spectrum are true horror stories of predators approaching minors, identity thefts, and the like. Most likely, you'll never deal with these issues, although we do touch on them at the end of this chapter.

On the other end of the spectrum are issues we usually categorize as "awkward social situations" — for example, posting a photo of your perfect beach day that your coworkers can see on a day when you called in sick (not that you would ever do something like that). You'll probably deal with issues at the awkward end of the spectrum most often. Somewhere in-between are questions about encountering hate speech, strangers seeing your stuff, and security issues such as spamming and phishing. All these privacy-related topics are legitimate, and all are ones you can learn how to deal with and control.

Regardless of where on the spectrum your question or problem falls, you should be able to use your privacy settings to make things better. We can't promise that you'll be able to prevent 100 percent of the situations that make you annoyed or uncomfortable or leave you with that sort of icky feeling, but we can tell you that you should be able to reduce how often you feel that way. The goal is to get as close to 100 percent as possible so you can feel as comfortable as possible sharing on Facebook.

Of course, there's also a spectrum of what being comfortable means to different people. That's why talking about privacy can get confusing: Lots of options are available, and what makes us comfortable might not make you comfortable. This chapter is a guide to all the privacy options Facebook offers so that you can figure out the right combinations that make you comfortable sharing on Facebook.

REMEMBER

Common sense will be one of the best helpers in avoiding privacy problems. Facebook status updates aren't the right place to post Social Security numbers or bank passwords. Similarly, if you're thinking about sharing something that could lead to bad real-world consequences if the wrong someone saw it, it's probably not meant to be shared on Facebook.

Knowing Your Audience

Before getting into specifics about privacy controls, you need to understand some words in the Facebook vocabulary. These terms are related to how Facebook thinks about the people you may or may not want to share with. For most pieces of information, privacy options are related to the audience who can see what you're sharing. The first two options in this list — Public and Friends — are the most basic settings and are shown by default whenever you go to change your privacy.

Figure 6-1 shows the menu used to change your audience for a post. You can choose from any of the following:

>> **Public:** By setting the visibility of something you post or list to Public, you're saying that you don't care who, on the entire Internet, knows this information about you. Many people list their spouse on their timeline, and, just as they'd shout this information from the treetops (and register it at the county courthouse), they set the visibility to Public. This setting is reasonable for innocuous pieces of information. In fact, some information, such your name, profile picture, cover photo, and gender, is always available as public information that everyone can see.

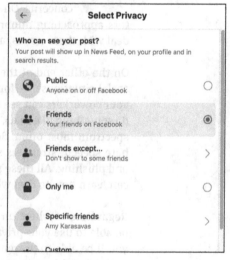

FIGURE 6-1:
Audience options for your posts.

Now, just because everyone *can* see something doesn't mean everyone *does* see everything. Your posts, information, friendships, and so on populate your friends' News Feeds (assuming that your friends can see this information), but never the News Feeds of people you're not friends with (unless you allow followers to see your public posts). When we think about who will see the information we share as Public, we imagine someone like you searching for us by name and coming to our timeline. Although (we hope) that might be a lot of people, it isn't anywhere close to the number of people who use Facebook. By default, much of your timeline and all your posts are publicly visible. This chapter covers how to change these settings if you want to.

>> **Friends:** Any information for which you set the visibility to Friends will be accessible by only your confirmed Facebook friends. If you trust your friends, this is a reasonably safe setting for most of your information. If you feel uncomfortable sharing your information with your friends, you can use the Custom privacy setting, or you can rethink the people you allow to become your friends.

Think of friending people as a privacy setting all its own. When you are about to add someone as a friend, ask yourself whether you're comfortable with that person seeing your posts and other information.

>> **Friends Except:** Even though we always recommend adding friends only if you're comfortable with them seeing your posts, we know that sometimes you have not-quite-friends on your friends list. These may be distant family members, professional contacts, old friends from way back when, or that super-friendly neighbor whom you wish wouldn't stop by quite so often. No matter who they are, you can exclude them from seeing your posts or other information by choosing the Friends Except setting. When you choose this setting, the screen shown in Figure 6-2 appears, listing your friends in alphabetical order by first name. You can use the search box at the top of the box to search for a specific friend. Select friends by clicking their names. This turns the circle next to their name red, making it look a bit like a Do Not Enter sign. Click the blue Save Changes button when you've finished excluding people.

The names you select are people who will *not* be able to see what you post. If you want to choose only the people who will see your posts, use the Specific Friends setting.

>> **Only Me:** This setting is a way to add something to Facebook but hide it from being seen by other people. At first, this setting may not seem useful, but it can come in handy for those times when you want to work on something, such as a photo album, and hide it from view until you finish it.

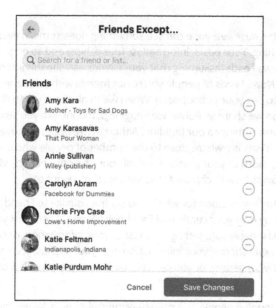

FIGURE 6-2:
Choose who you
don't want to see
something.

>> **Specific Friends:** This setting displays the Specific Friends dialog, where you select the people who *can* see something, as opposed to selecting those who can't see it. If you have something you want only a few people to see, this setting is a good way to share with that smaller group of people.

>> **Custom:** If you have specific needs, customized privacy settings may help you feel more comfortable sharing on Facebook. When you select Custom, the Custom Privacy window shown in Figure 6-3 appears. Customized privacy has two parts: those who can see something and those who can't.

In the top part of the Custom Privacy window, you enter the names of friends who you want to be able to see something. You can also type *friends of friends* to allow your post to be seen by your friends' friends. Your posts will be more visible than to just friends but not publicly available.

FIGURE 6-3:
The Custom Privacy dialog.

Additionally, the Friends of Tagged check box enables you to choose whether the friends of any people you tag can see your post. Remember, tagging is a way of marking who is in a photo, who is with you when you check in someplace, or whom you want to mention in a post. For example, say that you're going to spend a day at the park with one of your friends. You might post a status that says, "Taking advantage of the nice weather with **Marjorie.**" The name of your friend links back to her profile. By default, Marjorie's friends will be able to see this post, even if they aren't friends with you. If you deselect the Friends of Tagged option, Marjorie's friends will no longer be able to see that post.

The lower section of the Custom Privacy dialog controls who can't see something. Like the top part of the window, the Don't Share With section has a blank text box where you can type the names of people. When you add their names to this box, they won't be able to see the content you post.

REMEMBER

Whatever customized audience you create for one post will be the audience next time you go to post something. Make sure you check the audience the next time you post!

Changing Privacy as You Post

Although most people wind up selecting a privacy setting and sticking with it, thinking about your privacy settings on Facebook shouldn't be a one-time task. Because you're constantly using Facebook — adding status updates, photos, and content to Facebook, interacting with friend, and reaching out to people — managing your privacy is an ongoing affair. To that end, one of the most common places where you should know your privacy options is in the share box.

The *share box* is the blank text box that sits at the top of your Home page and under your cover photo on your timeline. It's where you go to add status updates, photos, links, and more to Facebook. The part of the share box that's important for this chapter is the privacy menu, which can be found below your name when you're creating a post.

Whenever you're posting a status or other content, the *audience*, or group of people you've given permission to see it, is displayed in the privacy menu. The audience displayed is always the audience you last shared something with. In other words, if you shared something with the public the last time you posted a status, the privacy menu displays Public the next time you go to post a status.

Click the Privacy button to reopen the Select Privacy dialog. Most of the time, we share our posts with friends. As a result, we don't change this setting that often. But if you do share something publicly, remember to adjust the audience the next time you post something.

WARNING

A post's privacy icon (Public, Friends, Friends Except, Only Me, Specific Friends, or Custom) is visible to anyone who can view that post. People can hover their cursor over that icon to get more information before they comment on or react to it.

After you post something, you can always change the privacy on it. From your timeline, follow these steps:

1. **Hover your mouse cursor over the privacy icon at the top of the post whose audience you want to change.**

Every post displays the icon for Public, Friends, or Custom.

2. **Click the privacy icon to open the Select Privacy dialog.**

You have the usual options: Public, Friends, Friends Except, Only Me, Specific Friends, and Custom.

3. **Click the audience you want.**

As soon as you click your new selection (or click the blue Save Changes button, for the more customized options), the privacy of that post changes and you return to your timeline.

Understanding Your Timeline Privacy

In addition to the content you post, you can control who can see the information you've entered in the About section of your timeline. This information, such as where you went to school or your relationship status, changes infrequently, if ever. You can edit the privacy for this content in the same place you edit the information itself. To get there, go to your timeline, and then click the About tab below your cover photo.

The About page has several sections, each representing a different information category. So, for example, all your work and education information appears in the Work and Education section. Next to each item is a privacy icon signifying who can see that piece of information. By default, most of this information is set to public and visible to everyone, although contact information is visible only to friends by default.

Clicking the privacy icon to the right of the field opens an interface for editing that information as well as changing who can see it. Use the privacy menu to change the privacy and click Save Changes when you're done. Rinse and repeat for any other pieces of information in the About section.

Timeline information is one of the places where the Only Me setting might come in handy. For example, lots of people don't like sharing their birthdays on Facebook, but Facebook requires you to enter a birthday when you sign up. By making your birthday visible only to you, you effectively hide it from everyone.

REMEMBER

Click Save Changes when you've finished editing privacy settings. Otherwise, the new settings won't stick.

Getting a Privacy Checkup

You can get to the Privacy Checkup page, shown in Figure 6-4, by clicking the Account icon (down arrow) in the top bar of any page, and then choosing Settings & Privacy ⇨ Privacy Checkup from the menu that opens.

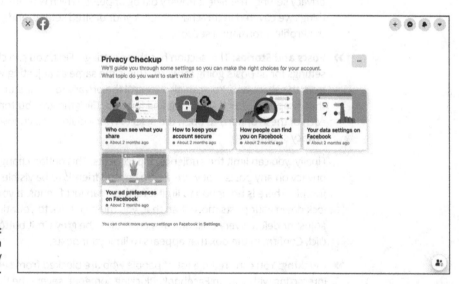

FIGURE 6-4:
The doctor is in for your privacy checkup.

On the Privacy Checkup page, you can not only learn about the different privacy settings but also change some of them. Click a topic to learn more about it and how it affects your privacy on Facebook. You can adjust settings as you go by clicking the grey button that displays your current setting for that information or topic. Doing so opens the Select Privacy dialog.

Who Can See What You Share topic

If you've been reading this book straight through, you've probably noticed that we mention privacy a *lot*. As we've said many times, you can change privacy for each piece of information you add to Facebook. What this means over time, however, is that when tasked with remembering who can see your hometown and who can see a post you made three weeks ago, you might have trouble finding the answer.

The Who Can See What You Share topic on the Privacy Checkup page helpfully collects all the pieces of information from your About section, your timeline, and even some parts of your profile you might not think of visible, and lets you see their current privacy setting. You can then adjust those settings if you want. When you click to open this section, a box appears that walks you through three different sections. Click the blue Next button to move between sections:

>> **Profile Information:** Profile information includes things such as your contact info, birthday, relationship, hometown, current city, education, and work history. Additionally, you find options here to choose who can see a list of your friends and who can see the people, Pages, and lists that you follow. By default, these last two options are set to public. To adjust the settings for any piece of information, click the grey button next to it to display its current privacy setting. The Select Privacy dialog appears, which was one of the first things we covered in this chapter. Click your desired setting, and you return to the Profile Information section.

>> **Posts and Stories:** This section has three settings. First, you can change your settings for all posts going forward. This is the same as adjusting your privacy from the share box. You can also control the privacy for your stories (posts that disappear after 24 hours) going forward. Click the grey button next to each of these options to open the Select Privacy dialog and choose your privacy setting.

Finally, you can limit the audience of past posts. This option changes the privacy on any posts that were visible beyond friends to be visible only to friends. There is no option to limit it further than just friends. If you're trying to lock down your posts more than that, you must go back to your timeline and adjust or delete every post one at a time. Click the grey Limit button and then click Confirm in the box that appears to limit your posts.

>> **Blocking:** You can create a list of people who are blocked from seeing or interacting with you on Facebook. Blocking someone severs the tie between your profiles (if one exists). When you block people, they can't see your profile information or posts. Click the blue plus sign and then type the name of the person you want to block.

After you have finished reviewing these settings, Facebook summarizes what you just looked at. Click the blue Review Another Topic button to return to the Privacy Checkup page.

How to Keep Your Account Secure topic

When we talk about your Facebook security as opposed to your Facebook privacy, we generally mean how easy or hard it would be for someone else to access and use your Facebook account. If someone else gets access to your account, they might do bad things like send out spam or try to scam your friends for money. These types of behaviors are scary but can be prevented by you. Performing a privacy checkup on your account security allows you to review the following factors in account security:

» **Passwords:** The most basic form of account security (anywhere on the Internet, not just on Facebook) is creating a strong, hard-to-guess password and sharing it with no one. Don't send it to friends and don't make it the same for every account. If you realize now that your password is not secure (no judgment) click the blue Change Password button to change it.

» **Two-Factor Authentication:** Two-factor authentication is an extra layer of protection that means a website needs two ways to verify that you are in fact the one logging in to your account. Similar to needing a driver's license and a utility bill to verify your home address, two-factor authentication means you will need to enter your password and supply a special one-time code to log in from a new device, browser, or location. To enroll in two-factor authentication, click the blue Get Started button, which will guide you through entering your mobile phone number so you can receive login codes when appropriate.

If you're worried that two-factor authentication sounds annoying, keep in mind that two-factor authentication will kick in only when you're logging in to Facebook from a new device or a new location. If you log in from the same computer every day, you won't have to wait for Facebook to text you a login code. If you receive a login code when you're not trying to log it, get thee to the Facebook help pages to report that your account has been compromised.

» **Alerts:** Whether or not you choose to turn on two-factor authentication, you can choose to receive alerts when someone logs into your Facebook account from an unfamiliar place. This means that if you travel across the country, or log in to Facebook from a friend's computer, Facebook will send you an alert. You can choose here how you want to receive this alert: You can get a Facebook notification, a message on Facebook, an email, or any combination of the three.

How People Can Find You on Facebook topic

Part of being on Facebook means that people can find you on Facebook. It can be disconcerting for some people to get a friend request from a stranger halfway around the world or a message from that same person that feels barely coherent. Performing a checkup on your search settings allows you, if you choose, to restrict who can find you and how they can get in touch with you:

>> **Friend Requests:** Friending people on Facebook is one of the most basic ways of connecting with them. By default, anyone who can find you on Facebook can send you a friend request. Right now, when all you're thinking about is random strangers finding you, the fact that anyone can friend you can feel scary! But keep in mind that you can't become friends with someone without sending and receiving friend requests. If you're getting too many friend requests from people you have no connection to, you can choose to receive friend requests from only people who are friends of your friends. Limiting friend requests to friends of friends will cut down on the number of people who can send you requests, while still allowing your friends' friends to add you.

>> **Phone Number and Email:** These settings are different than the contact info settings in the Who Can See What You Share topic. That topic was about who could *see* your contact information. This setting allows people who already have your email address or phone number to find you on Facebook using that information as a search term. For example, if a long-lost classmate is looking for you but you use a different name on Facebook, he or she might try searching your old school email in hopes that it makes a match with your profile. If this makes you uncomfortable, you can limit who can find you using this search method.

>> **Search Engines:** Search engines are sites such as Google or Bing that people use to search for, well, everything. By default, Facebook provides baseline public information about you so that if someone types your name in Google, your Facebook profile appears as one of the search results. If someone clicks that link, he or she will be able to see whatever information you have set as visible to Everyone on Facebook. For most people, this includes a profile picture, cover photo, and name. In general, Facebook doesn't make a distinction between someone logging in to Facebook and searching for David Rose or going to Google and searching for David Rose. If you, on the other hand, do see a distinction, you can turn off the setting (switch it from blue to grey) so that search engines can't link to your Facebook profile.

Your Data Settings on Facebook topic

The Your Data Settings on Facebook topic a bit of a catch-all category, with two sections:

» **Apps and Websites:** Chapter 15 is dedicated to explaining how apps, games, and websites can be enhanced when you choose to share your Facebook information with them. For now, let's leave it at understanding that there are ways of sharing your Facebook information with apps, such as websites you log into with your Facebook info or games you play with friends. When you use these apps, you agree to share some information from your Facebook account with them. This section isn't so much full of settings as it is a location where you can review the apps and games with which you've shared your Facebook data. You can revoke the permissions you've granted by clicking Remove next to any of the games, websites, or apps listed here.

» **Face Recognition:** By default, Facebook uses a facial recognition algorithm to suggest who to tag when people post photos. For example, when a friend uploads a cute photo of you, Facebook's algorithm looks for faces in the photo, matches the faces against other photos it has tagged of people on Facebook, and recommends that hey, maybe your friend should tag you. If you would prefer that Facebook not do this, turn off facial recognition by sliding the switch from blue to grey.

Your Ad Preferences on Facebook topic

Facebook is free, but the price you pay are the ads you see when you use the site. There is no way to fully turn off ads, but the Your Ad Preferences on Facebook section of the checkup page allows you to learn more and tweak your ad experience a little:

» **About Ads on Facebook:** Before you change any settings, Facebook wants to make sure you understand how ads work on Facebook. The main point here is that Facebook tries to serve you ads relevant to you. It figures out what is relevant by paying attention to the things you interact with and do on Facebook and other websites.

» **Profile Information:** In order to put relevant ads in front of you, Facebook allows advertisers to try to target their information based on general data. For example, a dating website wouldn't want to advertise to people who are already in a relationship. To be clear, when Facebook talks about letting advertisers create targeted ads, Facebook is not showing an advertiser your profile and saying, "Is this someone you want to advertise to?" Rather,

Facebook is saying, "We have this many thousands people who have a job title of Mr. Manager. Would you like to advertise to that group?" If you would rather not be targeted based on the data listed in this section, change these sliders from blue to grey.

REMEMBER

You will still see ads when you turn off targeting based on targeting information. The ads you see will simply be different than what you would see if you left these sliders on.

>> **Social Interactions:** One feature of Facebook ads is that Facebook tries to append to the ad itself information about your friends' interactions with an advertised product. For example, if a local coffee shop is advertising and promoting their Page, the name of friends who have already liked (or followed) that Page will appear next to the ad. This adds an implication of endorsement: "Hey, if Charlotte likes Kestrel Coffee, maybe I should check it out. She really knows her coffee shops." Facebook doesn't share if you've seen an ad or made a purchase from an ad. It shares if you've liked (or followed) a Page, RSVPed to an event, or taken another action that's already visible to your friends. If you'd rather these interactions not get appended to ads, change the setting, making these actions visible to Only Me instead of to your friends.

Now that you've finished your privacy checkup, pat yourself on the back. You have a better understanding of your privacy, who can see and contact you on Facebook, and how to keep your account safe.

Navigating the Settings Page

The privacy checkup is a good way to ease yourself into understanding privacy and the sorts of things you can control on Facebook. For even more detailed settings, you need to dive into the Settings page, which is something of a beast, as you can see in Figure 6-5. A ton of settings area available, ranging from regulating how frequently you get email notifications to adding credit cards to your Facebook Pay account to adjusting who can see your public posts. Not all of these settings are related to privacy, but many are.

The left side of the page lists the different sections of the Settings page. Click any section to view the settings and information about it. The left sidebar remains and the content on the right side of the page changes depending on which section you've opened, at least for most sections. We note any sections that take you to a different location on Facebook.

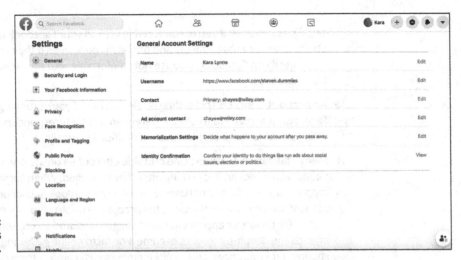

FIGURE 6-5:
The Settings
page.

REMEMBER

Sometimes other parts of the site will bring you to a specific section of the Settings page. You can figure out which section you're in by looking at the left sidebar and seeing which section is highlighted in grey.

General section

General Settings include basics such as your name and more arcane settings such as what should happen to your account after you die. Click the Edit link next to any option to expand the field. You can then learn more, change the information or settings, and click Save to save your choices:

» **Name:** By default, your name is the name you entered when you signed up for Facebook. You can come here to change your name to an alias or a new name, and you can also add other names (for example, a maiden name) that might help friends find you when searching. After you change your name, you can't change it again for 60 days, so make sure to double-check your spelling!

» **Username:** Your username is a personalized URL that you can use to get directly to your Facebook profile. People usually use this in a promotional capacity, the way you might add a URL to a personal website on a business card.

» **Contact:** The Contact setting displays the current contact email you use to log in to Facebook. If you would like to use additional email addresses or a mobile phone number to log in to Facebook, add them here. There is also a setting here to allow friends to download your contact info when they download information from Facebook. Later in the chapter, we provide more information about downloading info from Facebook, but for the purposes of this setting, ask yourself if you're comfortable with all your Facebook friends being able to see your email address and phone number, even if they aren't looking it up directly on Facebook. If you're not comfortable with this, shift the slider from blue to grey.

» **Ad Account Contact:** This setting is relevant only if you're running ads on Facebook. It allows you to use a different email for correspondence about ads than you use for your regular account notification emails.

» **Memorialization Settings:** One odd side effect of an online presence is that people's timeline can live on long after they have died. When someone is reported and confirmed to have passed away, Facebook *memorializes* their account, which is a way of locking the account — photo albums are hidden, the profile no longer appears in friend suggestions, and so on. Sometimes after death, people's accounts become a de facto online gathering site for sharing remembrances, and family members gain comfort from having a place where they can hear from their loved one's friends. Other times, family members find this legacy profile disturbing. Much like creating a will, you can decide whether you want to designate a legacy contact who can manage your account after you're gone or if you'd rather Facebook delete your account after you pass away.

» **Identity Confirmation:** Sometimes before you can run ads about certain topics, you need to verify your identity beyond the usual email confirmation. You begin this process here.

Security and Login section

If you jump back to the "Getting a Privacy Checkup" section, you may remember that account security is different than account privacy. Security has to do with how secure your account is from being used by other people. The Security and Login section displays information you can review for your own peace of mind, as well as settings you can change to protect your account:

» **Where You're Logged In:** This section provides a window into your most recent Facebook logins — what device, browser, and general location you logged in from. If you're the kind of person who uses the same computer and same phone to look at Facebook every day, the information here should look familiar. If you haven't been traveling and suddenly see a login pop up across the world, it may mean that someone got access to your account. Click the three dots icon to report that a particular login is not you. You will then be prompted to take steps to secure your account.

» **Change Password:** Your Facebook password should be unique to Facebook, hard to guess, and shared with no one. If any of those requirements are not met by your current password, go right now to this setting and click Edit. You will need to enter your current password before you enter your new password.

» **Save Your Login Info:** Now that we've impressed upon you the need to make sure no one else can use your account on your behalf, you can make it easier to log in and stay logged in. This option is best if you don't share your computer with anyone. It you to stay logged in on that particular browser and device. You will still need to log in when you go to use Facebook on a friend's computer. You may also be asked to log back in if you travel and are suddenly using your computer from a new physical location.

» **Use Two-Factor Authentication:** If you'd like extra security for your account, turn on two-factor authentication. When you log in from a new location or device, Facebook will require two forms of verification: your password and a second code sent to you via text or an authentication app that you can download to your phone. Turn on two-factor authentication by clicking Edit. Then choose if you want to use the app or text messages.

» **Authorized Logins:** Two-factor authentication applies only when you're logging in somewhere new or not currently authorized. You can review the locations where you've already authorized log-ins by clicking View next to this option. If you want to remove one of those locations (for example, if you logged in on a library computer and think you may have forgotten to log out), select the box next to the device you want removed and click Remove.

» **App Passwords:** You can use your Facebook account to log in to various apps. Then when you go to use those apps, you have to enter your Facebook login credentials. If you'd rather not enter these credentials with apps, you can choose to have Facebook generate special passwords for access to the apps, still connecting those apps with your Facebook account.

» **Get Alerts about Unrecognized Logins:** Whether or not two-factor authentication is turned on, Facebook can send you a notification when your account is accessed from a new device or location. Click Edit to decide whether to receive notifications and, if so, whether you want to receive them on Facebook, through Facebook Messenger, or by email.

» **Choose 3-5 Friends to Contact If You Get Locked Out:** Sometimes suspicious activity (such as sending out too many messages or posts) may get your account temporarily shut down. Too many messages is an indication that an account has been hacked and is now trying to get access to your friends' accounts or scam them out of money. Facebook simply locks accounts displaying these suspicious behaviors and waits for the accounts' owners to get in touch to verify that they are the only ones with account access. Should this happen to you, one way you can get verified is to designate three to five friends who can vouch for you being the owner of your account. Click Edit to enter the names of your friends.

>> **Encrypted Notification Emails:** We don't encrypt our emails and don't think it's a necessary security feature for most people, but realize that this setting is here in case you need it. If you want to your emails from Facebook encrypted, so that only you can decrypt them to read their contents, click Edit next to this option. You need to enter your OpenPGP Public Key to have encrypted emails. OpenPGP is a software program that you need to download and run on your computer to create a public key for yourself.

>> **Recover External Accounts:** If you've been using another service with your Facebook account and find yourself locked out, you can regain access with the saved account key found here.

>> **See Recent Emails from Facebook:** You can view a list of emails you've recently received from Facebook. This setting is handy because hackers and *phishers* — people trying to gain access to your account — often send you a fake email, with a link, saying your account needs attention. The link brings you to a site that looks like Facebook but isn't. When you enter your password to log in, the bad actors collect your info. If you receive an email from Facebook that looks suspicious, you can always come here (don't use the links in that email!) to double-check that it's real before taking any action.

TIP

If you received an email from Facebook asking for your password or any specific information such as your address or social security number, it is almost certainly a phishing attempt. Ignore or report the email.

Your Facebook Information section

To understand the section titled Your Facebook Information, it's important to understand what Facebook means by *information* in this context. For you as a user, there's a big difference in all the things you do on Facebook. Posting a photo is different from accepting a friend request, and both are different from creating a fundraiser. But from the perspective of the Facebook algorithms, each of these tasks is simply information. So anything you do on Facebook — joining a group, messaging a friend, adding a life event — is just a piece of information.

The Your Facebook Information section is where you go to learn how your information is being used and to take specific actions on your information at a whole:

>> **Access Your Information:** If you would like to take a deep dive into all the information Facebook has about you, click the View link next to this option to see it all, sorted by category. The Access page is shown in Figure 6-6. Tons of categories are here, and if you click into any one of them, you go to a version of Activity Log zeroed in on only that type of information, sorted by time.

FIGURE 6-6:
Access your
information here.

>> **Transfer a Copy of Your Photos or Videos:** Facebook enables you to copy and move your photos and videos to another online service or download them to your computer. Click the View link and enter your password to begin this process.

First, choose whether you want to transfer photos or video (you must repeat this process to transfer both). Second, choose from a menu of destinations, including Google Photos and Dropbox, and then click Next. You need to confirm the transfer with the intended recipient of the transfer. Your photos and videos will still exist on Facebook after the transfer is complete. If you want to remove them from Facebook, you must manually delete the albums or videos.

>> **Download Your Information:** Downloading your information isn't necessary most of the time for most people. However, if you find you do want to download your information, you can do so here. The information you see after you choose this option is the same information you were able to view in the Access Your Information section, except you can download a copy of it to your computer as a file in HTML or JSON format. You can choose the date range and information types for the information you want to download.

>> **Activity Log:** Activity Log is a way of looking at all your information based (mostly) on when it happened. You see all the actions you have taken (or that others have taken related to you) in one place. Activity Log has its own section later in this chapter.

>> **Off-Facebook Activity:** Off-Facebook activity is activity that happens on other websites or apps and is then shared back to Facebook by those sites and apps. Not all websites share information back to Facebook, but any apps or sites that you've linked to your Facebook account may share back information,

such as when you last logged in or visited their site. From the Off-Facebook Activity page, you can click the link to Manage Your Off-Facebook Activity, which lets you take a closer look at the sites and apps sharing information back to Facebook. You can also click the link to Clear History, which will erase this information from your Facebook account.

>> **Managing Your Information:** This option brings you to a form on Facebook's Help pages where you can get help with both Facebook and Instagram data, including help with changing the various privacy settings. If you are trying to change a specific setting, have read through this chapter, and still feel like you can't find the right place, this is a good place to try to get some outside help from Facebook.

>> **Deactivation and Deletion:** Deactivation and deletion are the two ways to remove your Facebook account. *Deactivation* is temporary and preserves all your Facebook data for when you're ready to log in again. *Deletion* is permanent and all your info gets erased. Check out the sidebar on deactivation and deletion to figure out which option is right for you.

DEACTIVATE OR DELETE?

Although we think Facebook is a great site that can be fun and rewarding to use, sometimes something changes. You might realize that you're spending too much time on Facebook, to the detriment of your closest relationships. Or you might find it's taking up too much mental or emotional energy, or not adding to your happiness. If this happens, you might be wondering whether you should delete your account or just deactivate it.

When you *deactivate* your account, it's sort of frozen. All the data remains on Facebook's servers, but people can no longer see your profile, find you in search, add you as a friend, and so on. You can continue to use Facebook Messenger when you deactivate. If you change your mind about using Facebook, you can always choose to log back in, which reactivates your account. You can then resume all your usual Facebook activities.

When you *delete* your account, it and all its information is permanently removed. Any photos you've added or posts you've made will be deleted from Facebook's servers. Your friendships will be removed. If you change your mind and rejoin Facebook, you will need to start from scratch. If you still think this is the right choice for you, we recommend downloading your information before you leave so you will have a copy of your photos before they are removed. When you go to the Delete Facebook page, you have a number of options you can choose before you click the Delete button, and Download Your Information is one of them.

Privacy section

If you've been reading this chapter straight through, the Privacy section of the Settings page should feel redundant. It offers you the option to change settings that we already reviewed during the privacy checkup.

The Privacy section is divided into three sections: Privacy Shortcuts, Your Activity, and How People Find and Contact You. (The Privacy Shortcuts section provides links to other parts of Facebook where you can adjust or learn about settings.) Next to most settings you can see both a bold word that displays the current setting, and a blue link to Edit that setting. For most settings, clicking Edit opens more detail about the setting and a drop-down menu where you can choose your desired privacy setting.

First, let's look at the Your Activity settings:

REMEMBER

>> **Who Can See Your Future Posts?:** As discussed in "Changing Privacy as You Post," you can change your audience or privacy for each post you make. Most people, however, don't change their privacy every time. Instead, they usually use a default setting when they share. That default is displayed here. Click the Edit link to change who can see these posts going forward.

This setting doesn't affect posts you've made in the past! If you want to change who can see past posts, skip down to the "Limit the audience" setting.

>> **Review All Your Posts and Things You're Tagged In:** This setting provides a link to Activity Log, which is covered in greater detail later in this chapter.

>> **Limit the Audience for Posts You've Shared with Friends of Friends or Public?:** If you'd like to change who can see posts you've made in the past, you can use this setting to change the setting on all your public posts and friends of friends posts to Friends. This doesn't change who can see posts that were already set to Friends. If you want to further limit who can see those posts, you must go through your timeline manually and change the setting for each post or remove people from your friends list.

>> **Who Can See the People, Pages and Lists You Follow?:** The Pages, people, and lists you follow are visible to the public by default. When people who are not your friend visit your timeline, they see your profile picture and cover photo and any other public info, including Pages and people you follow. You can change who can see this information by clicking the Edit link.

Now let's look into the settings for How People Find and Contact You. Remember, a lot of this information is covered in "Getting a Privacy Checkup," so you can always skip back to that section for a refresher on what it means for people to be able to find you or contact you on Facebook:

>> **Who Can Send You Friend Requests?:** By default, everyone on Facebook can send you a friend request. If you'd like only people who are already friends with your friends to be able to send you requests, click Edit and use the drop-down menu to change this setting to Friends of Friends.

>> **Who Can See Your Friends List?:** Use this setting to restrict who can see your list of friends on your profile.

>> **Who Can Look You Up Using the Email Address You Provided?:** This setting applies to people who are not your friends and are using search to find you. If this setting is set to Friends or lower, even people who know your email will not be able to find you using that information on Facebook.

>> **Who Can Look You Up Using the Phone Number You Provided?:** Same as the preceding setting, only applied to phone numbers instead of email addresses.

>> **Do You Want Search Engines Outside of Facebook to Link to Your Profile?:** By default, if someone types your name in a search engine such as Google or Bing, your Facebook profile appears in the results. If you'd prefer that this not occur, deselect the Allow. . . box that opens when you click Edit next to this setting.

I'M HAVING A PRIVACY FREAK-OUT; WHAT DO I DO?

We can't tell you how many times we've received a frantic email from a family member or friend saying something like, "Oh my gosh, my friend just told me that his friend was able to see these photos that I thought only my friends could see and now I'm freaking out that everyone can see everything. Do you know what to do?"

The first step is to take a deep breath.

After that, go to your timeline and click the View As icon (eye). Then you can click around your timeline as though you're someone who isn't your friend. If you think that person is seeing too much, change the Limit the Audience for Past Posts setting on the Privacy section of the Settings page. This setting pretty much changes anything that was visible to more than just friends to be visible to only friends. Then you can begin the process of adjusting your settings so that, going forward, you won't have any more freak-outs.

Face Recognition section

The next entry in Settings is Facial Recognition. This setting is covered in the "Getting a Privacy Checkup" section. In short, Facebook uses facial recognition software to recognize people in photos and recommend tags. If you don't like the idea of Facebook searching for you in photos, click Edit and use the drop-down menu that opens to change this setting from Yes to No.

Profile and Tagging section

The timeline, or profile, is basically where you collect all your stuff on Facebook, including posts (from you or from friends), photos, and application activity. (For details, see Chapter 5.) Your timeline allows you to look through your history and represent yourself to your friends.

Tags on Facebook are a way of labeling people in your stuff. For example, when uploading a photo, you can tag a friend in it. That tag becomes information that others can see as well as a link back to your friend's timeline. In addition, from your friend's timeline, people can get to that photo to see her smiling face. You can tag people and Pages in any type of post as well as in comments and stories. And just as you can tag friends, friends can tag you in their photos and posts.

The Profile and Tagging section of the Settings page allows you to control settings related to people interacting with you on your timeline and tagging you in posts. (To control the privacy on things *you* add to your timeline — that is, your posts — you use the privacy menu in the share box.) The Profile and Tagging Settings section is shown in Figure 6-7.

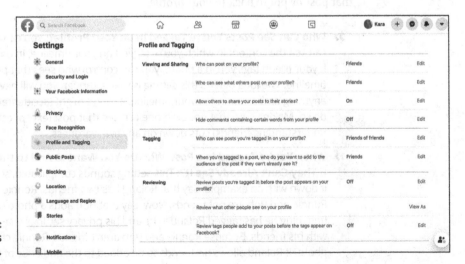

FIGURE 6-7:
Edit your settings
for tags.

>> **Who Can Post on Your Profile?:** By default, only friends can post on your profile. You can make it so no one (except you) can post on your profile using this setting.

>> **Who Can See What Others Post on Your Profile?:** Another option, if you're concerned about people seeing what other people have posted on your profile, is to change who can see this content. For example, if you don't want your work colleagues seeing references to your weekend exploits from your friends, you can make it so only certain friends (or only you) can see posts on your timeline.

>> **Allow Others to Share Your Posts to Their Stories?:** If you create public posts, by default people can reshare those posts as part of their stories (stories, you may recall, are a type of post that disappears from Facebook after 24 hours). If you'd rather your posts only live on your timeline, click Edit and then disable this setting.

>> **Hide Comments Containing Certain Words from Your Profile:** This setting is probably most relevant if you are a public figure or a sort-of-public figure and want to automatically maintain a certain tone or level of decorum on your page. After you click Edit, you can enter a list of words (or upload a .csv file) that won't be allowed on your timeline. This list might include swear words, phrases related to certain content, or emojis that you think might be offensive.

The second segment on the Profile and Tagging page relates to tagging. *Tagging* is the ability to label another person in a post. You might tag a photo or tag someone in the text of a post or comment. When someone tags you, it creates a link from that post or photo back to your profile.

>> **Who Can See Posts You're Tagged in on Your Profile?:** You can decide who can see the content in which you're tagged on your timeline. In other words, if your friend tags you in a photo, you can control who sees that photo on *your* timeline. The idea behind this setting is that, although you will never post anything embarrassing to your timeline, a friend might (accidentally, we hope) do so. Making sure that not everyone can see that post (except other, more understanding friends) cuts down on any awkwardness.

>> **When You're Tagged in a Post, Who Do You Want to Add to the Audience If They Can't Already See It?:** This setting sounds complicated, so let's break it down with an example. Say that Roland has two friends: Ronnie and Bob. Ronnie and Bob are not friends. Now, say that Bob adds a photo of Roland (meaning he has tagged Roland in it), and his privacy settings share that photo with his friends. Because Ronnie and Bob aren't friends, Ronnie can't see that photo. If Roland allows his friends to be added to the audience of that photo,

Ronnie *will* be able to see it. In other words, Roland can control whether Ronnie can see a photo of him that Bob added.

To change this setting, click Edit on the right side of the page and use the drop-down menu that appears to select who is added to the audience of a post you're tagged in. There are only three options for this setting: Friends, Only Me, and Custom. By default, this setting adds your friends to the audience of a post you're tagged in.

Finally, you can review certain posts or tags before they appear on your profile, if you'd like, with these settings related to reviewing:

» **Review Posts You're Tagged in Before the Post Appears on Your Profile:** Sometimes people don't like the idea of something being added to Facebook about them without their permission. What if someone adds an unflattering photo? While you can't stop someone from adding a photo, you can turn on this setting if you'd like to review the tag before it links back to your profile. In other words, your friend can still upload the photo and tag it, but you will receive a notification instead of the photo instantly appearing on your timeline as a photo you've been tagged in. The photo will appear on your timeline only after you've approved it. To change this setting, click Edit on the right side, and then use the drop-down menu to choose between enabled and disabled. By default, this setting is disabled (off).

» **Review What Other People See on Your Profile:** This isn't a setting but a tool that allows you to verify how your timeline appears to the outside world. This topic is covered in more detail in the "Checking out Facebook's privacy tools" section of this chapter.

» **Review Tags People Add to Your Own Posts before the Tags Appear on Facebook:** This setting controls tags your friends add to content you've uploaded. For example, if you upload a photo of 20 people to Facebook and don't tag anyone in it, your friends might choose to add tags. This setting lets you choose to review the tags your friends add before the tag is visible to other people.

To change this setting, click Edit on the right side of the page; then use the drop-down menu that appears to select whether Tag Review is enabled or disabled. By default it is disabled (off).

Public Posts section

Public posts, as you may recall, are the posts you make that are publicly visible. In other words, anyone using Facebook can see them. The settings in the Public Posts section pertain to only these public posts. If you never make public posts, you don't need to worry about this section at all. The settings here won't affect your friends' interactions with your posts that are set to Friends.

Many of the settings on this page might feel foreign if you mostly use Facebook to interact with people you know in real life. If you have a slightly more public existence, or use Facebook for outreach or as a promotional account, these settings may become more relevant:

>> **Who Can Follow Me:** Following, in this context, is like friending without the friendship. If people follows you, your public posts will appear in their News Feed. They won't be able to see your posts that are set to only friends, and you won't see their posts in your News Feed. You can allow anyone (Public) or only your friends to follow you. If you set this to Friends, you are preventing yourself from being able to use your Facebook account in a promotional or public way.

>> **Public Post Comments:** You can choose who is allowed to comment on your public posts. The three options are Public, Friends of Friends, and Friends. In other words, you can allow posts to be public, but keep any conversations they start just between friends or friends of friends. Keep in mind that comments on public posts would still be able to be viewed publicly.

>> **Public Post Notifications:** Use this setting to control which comments and reactions to your public posts generate notifications that you'll see on your Facebook account. If your public posts get widely distributed, the amount of comments and reactions can get high. Very high. If you've reached the point where you don't want to discourage the conversation but you want to stop notifications from cluttering your account, you can change this setting. You can restrict it so that comments from only Friends of Friends or Friends will create notifications, or turn off notifications entirely.

>> **Public Profile Info:** Parts of your profile that are always public, such as your name, profile pictures, and bio. Even though these parts are public, you can choose who can comment on or react to them here. You can choose to have Public, Friends of Friends, or Friends able to comment on this information.

>> **Off-Facebook Previews:** This setting has to do with Facebook posts that might be shared on other platforms (such as Twitter or Instagram). You can enable or disable previews of your posts to include your public information such as your profile picture even when not on Facebook.

>> **Comment Ranking:** For posts with lots of comments (again, if you are a public figure, comments could easily run into the hundreds or thousands), Facebook applies a comment-ranking algorithm that sorts comments by how relevant they are to the original post, as opposed to how recently the comment was made. You can turn off this algorithm if you would prefer all comments to appear in the order they were made.

>> **Username:** Use this setting to edit your public username, which is the same as your timeline address. This is the same setting found in the General section of the Settings page. Choose a username that is easy to type, is easy to remember, and relates to your authentic identity on Facebook.

Blocking section

Most of your privacy settings are preventive measures for making yourself comfortable on Facebook. Block lists are usually more reactive. If someone does something on Facebook that bothers you, you may choose to block him or block certain actions he takes from affecting you. You can manage seven block lists in the Blocking section: Restricted List, Block Users, Block Messages, Block App Invites, Block Event Invites, Block Apps, and Block Pages.

Using the restricted list

You can create a restricted list that accomplishes something similar to using the Friends Except privacy option. Friends who have been added to your restricted list can't see posts and other information visible to only friends. They will still be able to see public posts you've shared, and they'll see you interacting with people on Facebook. Think of a restricted list entry as one step below blocking the person.

To add someone to this list, click Edit List on the right side of the page. A pop-up window appears. If you've already added people to this list, they appear here, and you can remove them from the list by hovering your cursor over their pictures and clicking the X that appears.

To add people to the list, follow these steps:

1. **Click the On This List button in the top left.**

   A menu of two options appears: Friends and On This List.

2. **Click Friends.**

   A grid appears showing all your friends, alphabetically by first name, as shown in Figure 6-8.

3. **Select friends to add to the restricted list.**

   To select people, click their pictures, or search for them in the search box in the upper-right corner and then click their pictures.

4. **When you're done, click the Finish button in the lower-right corner of the box.**

Edit Restricted

Friends ▾   Search...

Amy Kara   Amy Karasavas   Annie Sullivan   Carolyn Abram   Cherie Frye Case

Katie Feltman   Katie Purdum Mohr   Linda Morris   Lindsay Sandman Lefevre   Mary Niemiec Bednarek

Paul Chen   Peter W. Knox   Steve Hayes   Teresa Bennett Mankin

Add and remove people from this list by clicking on their names.   Finish

FIGURE 6-8:
Add people to the
restricted list.

Blocking users

Blocking a person on Facebook is the strongest way to distance yourself from that person on Facebook. For the most part, if you add someone to your blocked list, he can't see any traces of you on Facebook. You won't show up in his News Feed; if he looks at a photo in which you're tagged, he may see you in the photo (that's unavoidable), but he won't see that your name has been tagged. When you write on other people's timelines, your posts are hidden from him. Here are a few key things to remember about blocking:

>> **Blocking is almost entirely reciprocal.** If you block someone, she is just as invisible to you as you are to her. So you can't access her timeline, nor can you see anything about her anywhere on the site. The only difference is that if you blocked the relationship, you're the only one who can unblock it.

>> **People you block are not notified that you blocked them.** Nor are they notified if you unblock them. If they're savvy Facebook users, they may notice your suspicious absence, but Facebook never tells them that they've been blocked by you.

>> **You can block people who are your friends or who are not your friends.** If you're friends with someone and then you block her, Facebook also removes the friendship. If you later unblock her, you will need to re-friend her.

REMEMBER

Blocking on Facebook doesn't necessarily extend to apps and games you use on Facebook and around the Internet. Contact the developers of the apps you use to learn how to block people in games and apps.

To add people to your blocked list, simply enter their names or email addresses into the boxes provided. Then click the Block button. Their names will appear in a list here. Click the Unblock link next to their names if you want to remove the block.

Blocking messages

Blocking someone from sending you messages is useful if the person is bothering you through either the Facebook Messenger app or by sending you Facebook messages from a computer. *Messages*, in this case, includes chats, voice, and video calls from the Facebook Messenger app. If someone is bothering you in comments sections or on your timeline, you need to fully block that person to get him to leave you alone. But if messages are the only problem, you have an easy solution here.

Note that you can add only friends to the messaging blocked list. If you're being bothered by someone who is not a friend, you need to report the person, or block the person entirely, or both.

To add people to the messages blocked list, simply enter their name in the box provided and press Enter. Their name appears in a list. Click the Unblock link next to their name if you want to remove the block.

Blocking app invites

An *app* is a term used to describe pieces of software that use Facebook data, even when those apps weren't built by Facebook. As friends use apps and games, they may send you an invite so you can join the fun. This is all well and good until certain people send you wayyyy too many invites. Rather than unfriend or block the overly friendly person who's sending you all those invitations, you can simply block invitations. You can still interact with your friend in every other way, but you won't receive application invites from him or her.

To block invites from a specific person, just type the person's name in the Block Invites From box and click Enter. That person's name will appear in the list below the text box. To remove the block, click Unblock next to that name.

Blocking event invites

As with app invites, you may have friends who are big planners and love to invite all their friends to their events. These may be events that you have no chance of attending because, say, they're taking place across the country and your friend has chosen to invite all his friends without any regard for location. Again, your friend is cool; his endless unnecessary invitations are not. Instead of getting rid of your friend, you can get rid of the invitations by entering his name here.

To block event invites from specific people, just type their names in the Block Invites From box and click Enter when you're done. Their names then appear in the list below the text box. To remove the block, click Unblock next to their names.

Blocking apps

Occasionally, an app behaves badly once you start using it. By *behave badly*, we mean the app spams your friends or uses your information in ways that make you uncomfortable. You can block the app to prevent it from contacting you through Facebook and getting updated Facebook information about you.

To block an app, type its name in the Block Apps text field and press Enter. The name of the app appears in the list below the text box. To remove the block, click Unblock next to its name.

Blocking Pages

Pages are basically timelines for non-people (businesses, brands, famous people, pets, anything that isn't covered by the term *person*). Posts from Pages may appear in your News Feed, either because you followed them, because they're advertising, or because Facebook's algorithms are recommending them to you. Pages can often interact in many of the same ways as regular people on Facebook (for example, commenting on or liking your posts), which means you might find yourself needing to block one. When you block a Page from interacting with you, it will no longer be able to interact with you or your posts, and you won't be able to interact with that Page via message or post. It also automatically includes un-liking and unfollowing the Page. Importantly, you won't see any more posts from that Page in your News Feed.

To block a Page, type its name in the Block Pages text field and click Enter. The name of the Page appears in the list below the text box. To remove the block, click Unblock next to its name.

Location section

If you use Facebook on your mobile device, you should know that Facebook, like many apps, will create a location history for you based on the places your phone goes. Although it won't share this info with anyone, it will use it to help you explore locations near you, and it may factor into what ads you see. Some people see this as a bonus — they get more relevant information in their News Feeds and in their ads. Others see this as an intrusion — it's not Facebook's business where they go. If you fall in the second camp, you can turn off the Location History setting.

Language and Region section

Chances are, if you start using Facebook in your home country, mostly stay in your home country, and mostly have friends in your home country, you'll never need to touch the settings on the Language and Region page. However, many Facebook users frequently cross borders and speak different languages. If you're one of them, Facebook can be a great way to keep in touch with friends and families who speak different languages. By default, Facebook's language settings will follow the language settings of the device it's on. So depending on how you use Facebook, even if you move abroad, you may never need to monkey with these settings. Also, these settings are unique to the device you're using. So if you switch between a computer purchased in the US and a cellphone purchased in Korea, you may have different settings on those two devices. By default, most of these settings will be set to the same language. In other words, if you use Facebook in English, Facebook assumes you want to see any translated content in English.

The settings here are as follows:

>> **Facebook Language:** First and foremost, this is where you come to adjust the languages in which you use Facebook. You want to be using Facebook in the language you read best. If you change this from English to Korean, for example, all menus, help pages, and buttons will appear in Korean instead of English.

>> **Formats for Dates, Times and Numbers:** When the date is 12-31-20 in the US, it's 31-12-20 or 20-12-31 across the pond (as ignorant Americans, we're not actually sure). You can choose a region from a drop-down menu and dates, times, and numbers will appear in regionally appropriate fashion.

>> **Temperature:** Similarly, you can toggle between Fahrenheit and Celsius. There aren't that many places on Facebook that display the temperature, but when it appears, it's nice to see it in the right unit.

>> **Language You'd Like to Have Posts Translated Into:** Some posts from friends and Pages may be translated by Facebook's automated translation system. You can choose the language you'd like to see posts translated into. Again, Facebook assumes that this language is the same as the language you use Facebook in.

>> **Languages You Don't Want to Be Offered Translations For:** When you are using Facebook, you may see a post or comment written in a foreign language, with See Translation link next to it. By default, this link doesn't appear when you're using Facebook in your designated Facebook language. If you don't want to see this link next to certain languages, because you can already read what they say, add those languages to this list.

>> **Languages You Don't Want Automatically Translated:** Similarly, some posts are automatically translated and displayed in your designated Facebook language. This is particularly true for Pages that have lots of fans across borders. If you would prefer to see posts in their original language, as opposed to their translated versions, add the languages this applies to here.

>> **Multilingual Posts:** This setting allows you to write a post in more than one language. This setting can be helpful, for example, if you want everyone in the world to see photos of your beautiful new baby, and want to add a caption in both Korean and English, for both sides of the family.

Stories section

A story is a special kind of Facebook post that disappears from Facebook 24 hours after it first appears. You can choose your privacy for your stories when you create them, just as you would for a regular post. However, in the Stories section of the Settings page, you have two additional settings regarding people resharing your stories:

>> **Allow Others to Share Your Public Stories to Their Own Story?:** If you create a story and share it with the public, you can decide whether or not people can add your story to their story. This allows your story to be seen by more people, but it also means it won't fully disappear from Facebook after 24 hours. Instead, it is still living a new life in someone else's story.

>> **Allow People to Share Your Stories If You Mention Them?:** Since you can tag or mention people in your stories, these tagged friends may want to reshare your story. You can allow them to do so or not. Again, if you allow them to reshare your story, it may have a longer life on Facebook than its original 24 hours.

Notifications section

Notifications are the little alerts you get on Facebook, in your email, or on your phone about things that happen on Facebook. Most notifications are about things that involve you, such as when someone tags you, adds you as a friend, or comments on your post. Depending on your settings, you might also see notifications about posts to a group you belong to, reminders about birthdays, and more. When you go to the Notifications section of the Settings page, you see the full list of categories for notifications you can receive, ranging from comments to people you may know to Pages you follow.

Click any category to view your settings for that category. Figure 6-9 shows the settings for Comments.

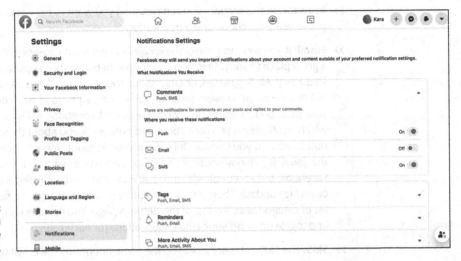

FIGURE 6-9:
Control how you
receive
notifications.

Each category on this page has three sliders you can turn on (blue) or off (grey):

» **Push:** Imagine Facebook as a tiny robot. Mostly this robot wants to get out of your way and doesn't talk to you directly. It might show you News Feed articles it thinks you'll care about, but it doesn't bother you about them. However, when someone comments on a post you made, it nudges you by popping up a little window in the lower-left corner of the page and by adding a red flag to your notifications icon. Facebook refers to these as *push notifications* because it's pushing something to your attention. If you don't want these notifications for a particular category, move the slider to off.

» **Email:** Facebook often sends email notifications to you in addition to displaying a notification on the website itself. If you spend a lot of time on Facebook, you might find this behavior redundant, in which case you can turn it off.

» **SMS:** You may also receive Facebook notifications as a text message on your phone. Keep in mind that if you have a smartphone and use the Facebook app on it, the SMS system is different. Receiving both SMS and phone notifications can get overwhelming. So if you use the Facebook app, we recommend that you keep this slider switched to off.

If you know you want to turn on or off a certain notification for most categories, you can sort the notifications by *how* you receive them. Scroll to the bottom of the page, to the How You Get Notifications section. Click any of the three categories to expand them and see all the options:

» **Browser:** In this section, you can choose if you want to hear sounds when you're using Facebook and a notification or message comes in.

» **Email:** If you want to control the emails you get from Facebook, this option is a great place to start. First, you can choose between three options for email frequency: All, Suggested, or Only about Your Account. If you choose All, you will receive an email every time someone posts a photo of you, comments on your post, and so on. If you choose Suggested, Facebook will try to determine which notifications you care about and only email you about those types of notifications. If you choose Only about Your Account, you will receive none of the usual day-to-day notifications about interactions with your friends on Facebook, but Facebook will still be able to email you regarding a security alert or privacy update. Then, if you continue to scroll down, you can see the same list of categories as the top of this Settings page. You can quickly go through and toggle on or off your email notifications for each category.

» **SMS:** The SMS category gives you the same options as the email category (assuming you have activated your phone on Facebook). You can view the mobile number where you will receive notifications, decide on your preferred frequency of SMS notifications (depending on your phone plan, you might be paying more if you receive more text messages). You can then browse through the list of notification categories and toggle on the ones you want to receive, and toggle off the ones you don't want to receive.

Mobile section

In the Mobile section of the Settings page, you can adjust settings for the phone number linked to your account as well as other details about text messages from Facebook. These settings are covered in Chapter 7.

Apps and Websites section

An *app* is a piece of software that uses Facebook data, even when the applications wasn't built by Facebook. An app might be a game, website, or tool, all of which make use of the data you already share on Facebook. To make it easier to use these applications, they import the data from Facebook. The App Settings section is where you go to view and edit how apps, games, and websites interact with your timeline. This section is covered in depth in Chapter 15.

Instant Games section

Instant games are games you can play with friends on Facebook or through Facebook Messenger. Like other apps and websites, they're sometimes built by developers other than Facebook, so when you start using them, Facebook shares some data with them on your behalf.

The Instant Games section shows you any games that are Active, which means games you've played recently (within the last 90 days) that still have access to your information. Click the View and Edit link next to any game to open an App Settings box. This box displays the information that has been shared with the game, as well as any settings applicable to that game (such as, Can This Game Send You Notifications?). Use the drop-down menus to change these settings, and click the blue Save button when you are finished.

If you remove a game, it will no longer receive information from Facebook, but it may still have access to any public information or information you previously shared with it. To remove a game, follow these steps:

1. **Select the check box to the right of the game you want to remove.**

   You can select more than one game at a time.

2. **Click the blue Remove button at the top of the list.**

   A window opens asking you to confirm that you want to remove the game.

3. **Select the grey box if you want to delete any posts, videos, events, or other content that the game has posted to your timeline.**

   Not all games post to your timeline, so this decision may be irrelevant.

4. **If you want to notify the gamemaker that you're removing the game from your Facebook account, leave the box next to the Cancel and Remove buttons selected.**

   If you don't want to notify them, deselect the box.

5. **Click the blue Remove button.**

   The connection between your account and the game is removed.

6. **To return to the Settings page, click the blue Done button.**

   The game you removed can now be found under the Removed tab of the Instant Games section of the Settings page, should you ever need to get to it for reference.

Business Integrations section

Business integrations are apps that work with Facebook business accounts: Pages and advertisers. If you're posting on Facebook professionally, a number of apps can help you get organized and increase engagement with your business or brand. The Business Integrations section of the Settings page is a list of all the apps you've used as your Page or other business entity. It works the same way as the Apps and Websites section, which is described in Chapter 15.

Ads section

Instead of charging users money, Facebook pays the bills by selling ads, which are then shown to you. So looking at ads is the way you pay for using Facebook.

When you click the Ads section of the Settings page, the Ad Preferences page appears with the Advertisers portion selected. You can view a list of advertisers you've seen recently, change the frequency with which you see certain topics, and check out some settings you can adjust. You can exert some influence on some aspects of the ads you see, but no matter what, you will still see ads.

Note that Facebook never gives advertisers personally identifying information about you. Advertisers get only *aggregated data,* which is data that has been compiled about many users without specifically naming any one user. So an advertiser never sees that you, personally, clicked that one ad for the upcoming season of *Bridgerton* upwards of ten times. All they know is that their ad received a certain number of clicks from women in the over-35 age range. Similarly, Facebook doesn't show an advertiser a picture of you and tell them that "this person lives in Seattle and likes coffee and has two children and is afraid of clowns." Instead, if an advertiser selects coffee as an interest for their desired audience, they're told how many people might see that ad.

The Ad Preferences page has three sections. These pages are deliberately dense and full of extra information because Facebook wants you to leave these settings untouched. However, it is your Facebook and we do our best to explain what you can (and can't) tailor to your liking.

Advertisers section

This section displays information about the advertisers you've seen recently. If one advertiser is feeling way too omnipresent in your Facebook life (it happens), click the Hide Ads button next to that advertiser to stop seeing ads from them.

Below the list of advertisers you've seen recently are two other sections you can view: advertisers you've previously hidden (you can undo those hides, if you so choose) and advertisers you've clicked. Without knowing very much about how Facebook's ad systems work, it has become obvious to us over time that after you click an ad, Facebook knows you're interested in that ad and then shows you even more from that advertiser. If you need to put something out of sight and out of mind, you can hide ads from a certain advertiser even after you've clicked their ad.

Ad Topics section

The Ad Topics section gives you general categories of ads for which you can turn down the volume: Alcohol; Parenting; Pets; and Social Issues, Elections or Politics. These topics are ones that people complain about the most, and can be most upsetting to someone, say, going through a pregnancy loss or trying to get sober. Click the See Fewer button next to the topic you want to see less of. Click the Undo button to return to a normal volume of ads about that topic.

Ad Settings section

The top section of the Ad Settings section, showing in Figure 6-10, displays some common FAQs about how ads work on Facebook. Following are five categories of ad data you can adjust. Each option opens a pop-up box when you click it, providing you with more information about how that type of advertising works and, if you try hard enough, settings you can adjust:

>> **Data about Your Activity from Partners:** In general, Facebook shows each user personalized ads. To personalize those ads, it uses information you've added to Facebook as well as information about you that advertisers share back to Facebook through things such as cookies, which track your behavior online. You can turn off this personalization through partners by moving the slider next to your name to off. Remember, you will still see ads; they just won't be generated using data shared from partners and advertisers. If you have a linked Instagram account, you can also turn off these types of ads there.

>> **Categories Used to Reach You:** Advertisers can create a profile for the type of person that they want to see their ad. If you fit that profile, you'll see the ad. You can prevent yourself from being added to advertising profiles based on Employer, Job Title, Education, and Relationship Status. Use the sliders to turn off any categories you do not want used to target you by advertisers. Additionally, you can click through to view your Interest Categories and Other Categories.

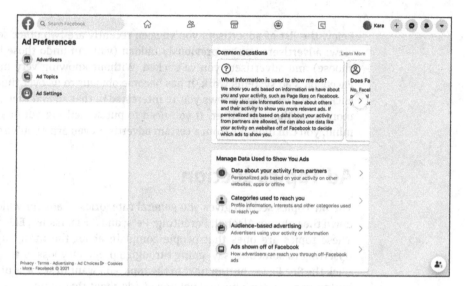

FIGURE 6-10:
Keep tabs on the ads you are seeing on Facebook.

Interest categories are the ones advertisers may use to find eyeballs, and they're the ones that Facebook has algorithmically attached to your profile. Some of them may seem obvious — if you liked the *Schitt's Creek* Page, *Schitt's Creek* will be one of your Interest categories. But others might not be clear; for example, multiple sclerosis awareness, though certainly a worthy cause, entered one of our interest categories. You can click remove next to any category to prevent advertisers from reaching you through that particular category. You may still see the same ads through other categories or other Facebook systems, but you won't see them if an advertiser is targeting only through that category.

The Other Categories that Facebook has here refers to categories that are based not on interests but on other data. For example, a jewelry company might want to target people who have an anniversary coming up, so one of your Other Categories might be Anniversary within 30 days. You can remove any category in this list by clicking the Remove button next to its name.

>> **Audience-based Advertising:** Audience-based advertising uses lists of information provided by advertisers that then get matched with Facebook's own information. For example, a company might upload an email distribution list, and then Facebook will show ads to people who are already on the company's mailing list (or, for that matter, maybe show ads to only people not yet on the mailing list). Click any company in this pop-up box to view more about the company and its Facebook presence. If you click through to the section labeled Why Are You Included in This Advertiser's Audience?, you can see the List Usage box. Here you can choose not to allow companies to show you ads or exclude you from seeing ads based on the lists they upload. Click the Don't Allow button to change these settings.

- >> **Ads Shown Off of Facebook:** Advertisers and Facebook share information back and forth. Even when you're using another website, Facebook may be the one showing you ads. You can decide whether advertisers can reach you through ads that are not on Facebook but generated by Facebook. To turn this setting off, move the slider from blue to grey.

- >> **Social Interactions:** This section should look familiar from the "Getting a Privacy Checkup" section, where we talk about ads in News Feed. Ads sometimes show social data — for example, the fact that you liked (or followed) a page, or RSVPed to an event — alongside the ad itself. You can choose whether friends can see your social data in their ads. If you would prefer that they not see this data, change this setting from Friends to Only Me.

    Even if you don't change this setting, only your friends ever see information about you next to ads. If we're not friends, you'll never see a notice that "Carolyn Abram likes this movie" or "Amy Kasaravas likes this grocery store" next to an advertisement for a product or company.

REMEMBER

Ads Payments section

The Ads Payments section brings you to Ads Manager, where you can adjust your payment settings if you are someone who pays to run ads on Facebook. If you're looking for help with your Facebook Page, ads, or business, check out another Dummies book, such as *Social Media Marketing For Dummies*, 4th Edition.

Facebook Pay section

Facebook allows you to send and receive payments between you and your friends. To do this, you need to enter credit card or PayPal info. You can get started using Facebook Pay by clicking the Facebook Pay section, which will bring you to the Facebook Pay page. Review recent payment activity, adjust your settings, and get help with your Facebook Pay account.

Support Inbox section

When you report inappropriate content to Facebook or get reported for inappropriate content, Facebook communicates with you through a separate support inbox that doesn't mix with your usual messages from friends. Clicking the Support Inbox section displays a page where you can review reports and how Facebook responded. You can also find additional links to Facebook's Safety Center and other safety-related information.

Videos section

Settings related to video quality and playback are in the Videos section. Many of these settings are set to default, which means that Facebook may change what it shows you depending on other settings. For example, if you are watching one video with the sound on, the next video you see will automatically have the sound on. You can make things always happen one way or another by adjusting the settings. These settings are as follows:

>> **Video Default Quality:** You can change this setting to be always in SD or to show in HD whenever it is available.

>> **Auto-Play Videos:** *Auto-Play* means the video starts playing as soon as you see it. If auto-play is off, you need to click a play button before a video begins.

>> **Always Show Captions:** You can turn on and off captions for videos.

>> **Captions Display:** You can adjust the appearance of captions to better suit your needs. For example, you can change the text's color or size.

>> **Video Default Volume:** If you select Quiet as the default volume, videos will always start at a low volume.

Understanding Privacy Shortcuts

To get to the Privacy Shortcuts page, click the Account icon (down arrow) in the upper-right corner. Click Settings & Privacy from the menu that opens, then click Privacy Shortcuts from the next menu. The Privacy Shortcuts page shown in Figure 6-11 appears.

For the most part, clicking any of the links on this page will bring you to a different page or a different section of Facebook. These are the *shortcuts* to solving your privacy problems and answering your privacy questions. The shortcuts in this section should feel familiar if you read the "Getting a Privacy Checkup" and "Navigating the Settings Page" sections. If you didn't, you may want to read those sections to discover how Facebook thinks about your privacy.

TIP

When you have a specific question (for example, "How do I change my password?"), start with the Shortcuts page. When you more generally want to know what the deal is with your privacy these days, go to the Privacy Checkup page.

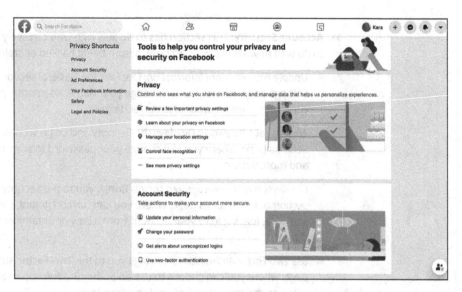

FIGURE 6-11:
Privacy shortcuts
can be found
here.

Following is a brief description of each section and its shortcuts on the Privacy Shortcuts page:

>> **Privacy:** This section has to do with the most common questions about privacy — who can see what about you. It contains the following shortcuts:

- *Review a Few Important Privacy Settings:* Returns to the Privacy Checkup page.

- *Learn about Your Privacy on Facebook:* Displays Facebook's educational pages describing how privacy works.

- *Manage Your Location Settings:* Brings you to the Location section of the Settings page, where you can turn location history on or off. This setting allows Facebook to build a history of where you've been based on where your phone or laptop has been.

- *Control Face Recognition:* Brings you to the Face Recognition section of the Settings page, where you can turn the Facial Recognition section on or off. You can read more about how this setting is used in the "Getting a Privacy Checkup" section.

- *See More Privacy Settings:* Displays the Privacy section of the Settings page, where you can review and change settings related to who can see your activity on Facebook and can find and contact you on Facebook.

- **Account Security:** This section has to do less with who can see you and more to do with how secure your account is from being hacked or stolen:

  - *Update Your Personal Information:* Displays the General section of the Settings page, where you can change your name, username, primary contact info, and memorialization settings.

  - *Change Your Password:* Displays the Security and Login section of the Settings page, where you can change your password to something new and more secure.

  - *Get Alerts about Unrecognized Logins:* Brings you to the Security and Login section of the Settings page, where you can turn on getting alerts about someone logging in to your account from a new or unfamiliar location or device.

  - *Use Two-Factor Authentication:* Brings you to the Two-Factor Authentication page, where you can choose to use an authentication app or SMS messages as the second way of verifying your login.

- **Ad Preferences:** This section has to do with the ads you see on Facebook:

  - *Learn about Ads:* Brings you to educational pages created by Facebook that attempt to explain how the ad system works and how your activity on Facebook and across the web relates to the ads you see.

  - *Review Your Ad Preferences:* Displays the Ad Preferences page, where you can review the advertisers and ad topics you've seen most recently and remove some or all these categories from your ads.

  - *See Your Ad Settings:* Displays the Ad Settings section of the Ad Preferences page, where you can manage the data that Facebook uses to show you ads.

- **Your Facebook Information:** This section allows you to review everything — and we mean everything — that you have done, shared, posted, or added on Facebook:

  - *Access Your Information:* Displays a list of your information categorized by type, such as posts, events, likes, and reactions. You can click a particular type to see a version of the Activity Log with just that type of information. Activity Log is covered in greater detail later in this chapter.

  - *See Your Activity Log:* Activity Log is a way of reviewing absolutely everything you've done on Facebook — posting something, joining a group, commenting on a post, adding a friend, and so on. More details on Activity Log later in this chapter.

  - *Manage Your Information:* Displays a form on Facebook's help pages where you can get help with both Facebook and Instagram data, including help with changing the various privacy settings. If you're trying to change a specific setting, have read through this chapter, and still feel like you can't find the right place, this is a good place to start trying to get some outside help from Facebook.

- *View or Clear Your Off-Facebook Activity:* Displays a page where you can review the apps and website that have shared activity with Facebook. You can view that activity in minute detail and clear the history.

  - *Delete Your Account and Information:* Displays a page where you can delete your account. See the "Deactivate or delete?" sidebar for more information.

» **Safety:** This section provides links to Facebook's educational pages, which are geared to helping you understand Facebook's policies and tools for keeping Facebook a safe and trusted place:

  - *Visit the Safety Center:* Displays Safety Center, where you can browse resources related to safety on Facebook.

  - *Find Resources for Parents:* Brings you to the Parents Portal section of the safety pages, where you can learn about keeping your kids safe online and on Facebook.

  - *Help Prevent Bullying:* Displays the Bullying Prevention Hub, which is a section of the Safety Center. Online bullying is a huge issue for many people, and you can learn more here about how to prevent it and what to do when you see it.

» **Legal and Policies:** Links to these legal documents are found as tiny links on your Home page. These are the legal documents you agreed to when you signed up for Facebook, including the terms of service, the data policy, the cookies policy, and community standards. If you want to review any of them, simply click them here.

Checking out Facebook's privacy tools

Okay, that was a lot of settings and a lot of information. What if you don't want to worry about these small settings and who tagged what when? What if you just want to make sure that your timeline looks the way you want to your friends and that people who aren't your friends can't see anything you don't want them to see? Well, the good news is that the View As tool allows you to do just that, and the Activity Log tool allows you to keep track of everything that's happened recently and to make adjustments without trying to figure out which setting, exactly, needs to be changed.

View As tool

To get to the View As tool, go to your timeline. Below your cover photo and name are several icons: Click the one with an eyeball icon to open the same page but with the View As tool active. The black bar running across the top of the page lets you know that you're currently viewing content on your timeline that is Public. You can click through to the various sections of your timeline. (Most people like to check, double-check, and triple-check the photos section.)

Note that no matter how much you've hidden your information and posts, everyone can see your cover photo and profile picture, and current city. Anything else the public can see can be hidden, if you so choose.

Activity Log

As you've probably noticed, a lot happens on Facebook. You take all sorts of actions: liking, commenting, posting, and so on. And people take all sorts of actions that affect you: posting on your timeline, tagging you in photos, and inviting you to join groups. If you want to know, line by line, everything that could possibly be seen about you by someone on Facebook, Activity Log is for you.

You can get to Activity Log from a few places. The simplest way is to open the Account icon (down arrow) at the top of the screen. Click Settings & Privacy from the menu that opens, and then click Activity Log to display the screen shown in Figure 6-12.

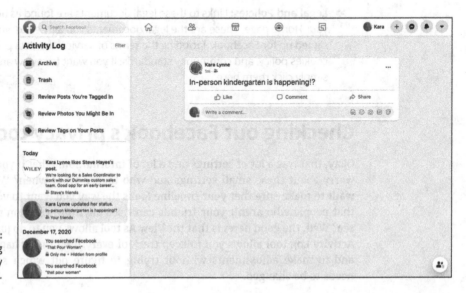

FIGURE 6-12:
Here's everything you did recently on Facebook.

When you're looking at Activity Log, note the sidebar on the left for navigating and the main section on the right for viewing posts or content associated with an item from your Activity Log.

Activity Log runs in reverse chronological order, so actions you've taken most recently are at the top; scroll down to see more. You can click the blue Filter link at the top of the left sidebar to view only posts of a certain type (such as only photos, or only likes and reactions). At the top of the left section are five sections you can jump to: Archive, Trash, Review Posts You're Tagged In, Review Photos You Might Be In, and Review Tags on Your Posts. If you have any review settings turned on, you will come here (usually by clicking a Facebook notification) whenever you have new posts, photos, or tags to review.

When you're looking at an individual line item, you see several pieces of information. First is an icon and a sentence explaining what you did (or what a friend did), something like "Carolyn wrote on Dana's timeline" or "Carolyn was tagged in Dana's photo." Facebook includes private actions such as your searches as items in your log. Each item has a preview of that post, photo, comment, or whatever it is related to. For example, if you commented on a photo, the preview will display the photo and your comment.

Below the preview is an icon and text representing who can see that item. This text might consist of the usual privacy options, or members of a group you belong to, or in the case of a post to a friend's timeline, that person's friends.

Clicking a preview displays the full content in the main section of the page. If it's a post, you'll be able to interact with it just as you would if you encountered it on a timeline or your News Feed. You can click links to edit it, edit the privacy for it (if it's your post), comment, like, and so on. To change the privacy on a post, click the icon representing its current privacy level. The Privacy window appears, and you can select a new setting. Remember, though, that for many of the items in Activity Log, you won't be able to change the privacy. For example, you can't change the audience for a comment on someone else's post. If you realize a comment you made or something you liked is visible to more people than you want, your only option is to edit or delete that content.

REMEMBER

When we say something is visible on timeline, we also mean that your friends might see that item in their News Feed.

You can see options for each item in your log by clicking the three dots icon that appears when you hover your mouse cursor over the item. An item could have a variety of options, as follows:

>> **Delete:** Many types of posts have only this option. Deleting something from Activity Log erases that content from Facebook. Keep in mind that you can

delete only things on your timeline or that you've created. If you delete a comment you made on a friend's post, you are deleting only your comment, not your friend's post.

>> **Hide from Profile:** Hiding something from your profile means it won't appear on your timeline, but it will still exist on Facebook. For example, if you create an event but hide it from your profile, the event still exists, people will still be able to RSVP, but there will be no reference to it on your timeline.

>> **Move to Archive:** The archive is a separate section of Activity Log where you can basically stash posts or other content that you don't want to delete but also don't want to have active on Facebook. If you are a digital hoarder who just hates the permanence of deleting something (what if you look back on that with joy someday?), having an archive for posts is like having a shoebox in the back of the closet for old photos.

>> **Move to Trash:** Trash is a separate section of Activity Log. Anything you put there is deleted after 30 days.

>> **Unlike:** Likes are common entries in Activity Log. If you want to remove a like, simply unlike it.

You can view anything that you move to the archive or trash by navigating to the Archive or Trash section, respectively, in the left sidebar of Activity Log.

We find Activity Log useful in that it helps us understand all the ways we participate on Facebook and all the things our friends might see about us and our life. But we've found that we don't change the privacy or the timeline settings for items here all that often.

Remembering that it takes a village to raise a Facebook

Another way in which you (and every member of Facebook) contribute to keeping Facebook a safe place is in the reports you submit about spam, harassment, inappropriate content, and fake timelines. Almost every piece of content on Facebook can be reported. Sometimes you may need to click the Options link or the three dots icon to find the report link.

Figure 6-13 shows an example of someone reporting an inappropriate photo.

The reporting options vary depending on what you're reporting (a group as opposed to a photo, for example). These reports are submitted to the Facebook Help Team, which then investigates and takes down inappropriate photos, disables fake accounts, and generally strives to keep Facebook safe and inoffensive.

When you see content that you don't like — for example, an offensive group name, hate speech, or a vulgar timeline — don't hesitate to report it. With the entire Facebook population working to keep Facebook free of badness, we can prevent a lot of inappropriate content from disseminating too widely.

FIGURE 6-13:
Reporting inappropriate content.

REMEMBER

After you report something, Facebook's Help Team evaluates it in terms of whether it violates Facebook's terms and community standards. This means that pornography gets taken down, fake timelines are disabled, and people who send spam may receive a warning or even have their account disabled. However, sometimes something that you report may be offensive to you but doesn't violate community standards and, therefore, will remain on Facebook. Due to privacy restrictions, the Help Team may not always notify you about actions taken because of your support, but rest assured that the team handles every report. Facebook will correspond with you about reports you make in the Support Inbox, which you can find in the Quick Help menu.

Peeking Behind the Scenes

Facebook's part in keeping everyone safe requires a lot of manpower and technology power. The manpower involves responding to the reports that you and the rest of Facebook submit, as well as proactively going into Facebook and getting rid of content that violates community standards.

The technology power that we talk about is kept vague on purpose. We hope that you never think twice about the things that are happening behind the scenes to protect you from harassment, spam, and pornography. Moreover, we hope you're never harassed or spammed, or accidentally happen upon a pornographic photo. But just so you know that Facebook is actively thinking about user safety and privacy, we present a few of the general areas where Facebook does a lot of preventive work.

Protecting minors

In general, people under the age of 18 have special visibility and privacy rules applied to them. For example, their profiles usually don't appear in public search listings the way that adults' profiles do.

Other proprietary systems are alerted if a person is interacting with the timelines of minors in ways they shouldn't or targets an ad to minors. Facebook tries to prevent whatever it can, but this is where some common sense on the part of teens (and their parents) can go a long way toward preventing bad situations.

REMEMBER

You must be at least 13 years old to join Facebook.

Preventing spam and viruses

Everyone can agree that spam is one of the worst parts of the Internet, all too often sliming its way through the cracks into email and websites — and Facebook as well, sometimes in the form of messages to you, timeline posts, groups, or events masking as something they're not to capture your precious attention.

When you report a piece of content on Facebook, "It's spam" is usually one of the reasons you can give for reporting it. These spam reports are incredibly helpful. Facebook has a bunch of systems that keep track of the sort of behavior that spammers tend to do, such as message people too quickly, friend too many people, or post a similar link in too many places. If you end up really taking to this Facebook thing, at some point you may get hit with a warning to slow down your messaging. Don't take it personally, and just follow the instructions in the warning — this is the spam system at work.

Preventing phishing

Phishing refers to malicious websites attempting to gain sensitive information (such as usernames and passwords to online accounts) by masquerading as the sites you use and trust. Phishing is usually part of spamming: A malicious site acquires someone's Facebook credentials and then messages that user's friends with a link to a phishing site that looks like Facebook and asks them to log in. They do so, and now the bad guys have a bunch of new Facebook logins and passwords. It's a bad cycle. The worst part is that many of these Facebook users get locked out of their own accounts and are unable to stop the spam.

Facebook has a series of proprietary systems in place to try to break this cycle, just like it does with spam and viruses. If you have the misfortune to get phished (and it happens to the best of us), you may run into one of the systems that Facebook uses to help people take back their timelines and protect themselves from phishing in the future.

The best way to protect yourself from phishing is to get used to the times and places Facebook asks for your password. If you just clicked a link in Facebook and suddenly a blue screen asks for your information, be suspicious! Similarly, remember that Facebook will never ask you to email your password to them. If you receive an email asking for something like that, report it as spam immediately. Also, beware when friends send attachments that you must download or strange messages from friends who don't normally message you.

TIP

If you want to stay up-to-date with the latest scams on Facebook or want more information about protecting yourself, you can like Facebook's Security page at `www.facebook.com/security`. Doing so will provide you with ongoing information about safety and security on Facebook.

One Final Call to Use Your Common Sense

No one wants anything bad to happen to you because of something you do on Facebook. Facebook doesn't want that. You don't want that. Your authors definitely don't want that, and we hope that these explanations help to prevent anything bad from happening to you on Facebook. But no matter what, *you* need to take part in keeping yourself safe. To ensure your own safety on Facebook, you have to make an effort to be smart and safe online.

So what *is* your part? Your part is to be aware of what you're putting online and on Facebook. You need to choose whether displaying any given piece of information on Facebook is risky. If it's risky, you need to figure out the correct privacy settings for showing this information to the people you want to see it — and not to the people you don't.

REMEMBER

Your part is equivalent to the part you play in everyday life to keep yourself safe: You know which alleys not to walk down at night, when to buckle your seatbelt, when to lock the front door, and when to toss the moldy bread before making a sandwich. Now that you know all about Facebook's privacy settings, you also know when to use the various privacy options and when to simply refrain from posting.

Chapter **7**

Facebook on the Go

Writing this chapter began with an existential question: Does this chapter need to exist? Not because you'll never use mobile features — the opposite. Using Facebook on a mobile phone is virtually indistinguishable from using it on the computer.

In part, that's by design: The Facebook app is designed to look and feel like Facebook on your computer. Any actions you take on Facebook on the web are visible on Facebook on your phone and vice versa. You can start a comment thread on your phone and pick it up on your computer. You can post a photo from your computer and check how many likes it received while you're on your phone at a coffee shop later.

Many people use Facebook more from their phones than from their computers. When you're out and about, you see things happening that you want to share. You take photos on your phone and want to get them out to Facebook-world immediately. When you're out and about is also when you tend to have time to kill: waiting for a doctor's appointment to start, sitting on a bus, pacing back and forth with a screaming baby. At these times you might want to read about your friends, catch up on interesting links and articles, and generally know what's going on around you.

In this chapter, we make a few foolish assumptions: that you have a iPhone or Android phone, know how to use its features and apps, and can browse the web from your mobile browser. If you have a mobile phone that can send and receive text messages but can't use apps or browse the web, skip to the last section, "Facebook Texts."

The Facebook App

The *Facebook app* exists for iPhones and Android phones. Carolyn is an iPhone user, so the steps and figures in this section are based on her experiences using Facebook for iPhone. If you're an Android user, worry not; the functionality is largely the same, and many of the steps will likely be similar to what you see on your phone. The biggest difference will be the location of certain buttons and links.

Before you can get started with the Facebook app, you need to download the app to your mobile phone: Android users from Google Play, and iPhone users from the App Store. The app is free to download and use. After you download the app, log in to your Facebook account (shown in Figure 7-1). You need to log in the first time you use the Facebook app; after that, your phone remembers your login info.

REMEMBER

Even after you download and log in to the Facebook app, you may need to additionally grant it access to things like your phone's camera roll, camera, and contacts to use all Facebook's features in the app.

FIGURE 7-1:
Log in to the Facebook app for iPhone.

Layout and navigation

The Facebook app's Home screen (shown in Figure 7-2) has many of the same features as the Home page on the web. Icons appear at the top and bottom of the screen. The top bar shows the Facebook logo, a search icon, and a Messenger icon. These disappear or changes as you scroll down the page and when you are in different parts of the Facebook app.

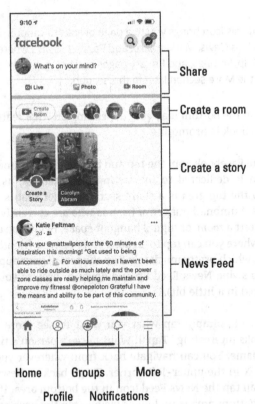

— Share

— Create a room

— Create a story

— News Feed

FIGURE 7-2:
Facebook for
iPhone home.

Home   Groups   More

Profile   Notifications

The bottom bar behaves more like the Facebook website's big bar on top. Even as you scroll and navigate the app, the bottom bar remains in place. It features five icons (labeled in Figure 7-2):

>> **Home:** Tap this icon at any time to return or reload your Home page. The screen automatically scrolls up to the top of News Feed, so if you've been scrolling for a while, this can be a quick way back to the share box.

>> **Profile:** Tap this icon to go to your own profile.

>> **Groups:** Tap this icon to view a feed of new posts to the groups you belong to. For more on using Facebook Groups, see Chapter 10.

>> **Notifications:** A small red bubble will appear at the top of this icon when you have new notifications about things that have happened on Facebook: for example, when someone has commented or liked a post of yours, responded to a comment you made, or added you as a friend. Tap this icon to view old and new notifications alike.

>> **More:** Tapping this icon brings you to a page of links to various parts of Facebook, such as News, Marketplace, and Pages. It is also where you can find links to Help, Settings, and Privacy pages and to log out. We provide details about the More section later in this section.

The icons you see at the bottom may vary depending on what features you use and what features Facebook is promoting.

When you open the Facebook app, the top and bottom areas frame the screen. The rest of the screen is dedicated to interacting on Facebook, as you can see in Figure 7-2. Below the top area is a share section. Under that is the create room area, which shows thumbnail images of friends who are currently online and with whom you could start a *room*, or digital hangout space. Below the create room area is the story area, where you can tap to create a story or view friends' stories. Below that is News Feed, which pretty much takes up the rest of your app's Home screen. Use your finger to swipe News Feed up and down to browse your friends' posts (more on News Feed in a little bit).

To navigate Facebook, simply tap what you want to see more of. See a photo thumbnail that looks interesting? Tap it. Want to see a person's timeline? Tap the person's face or name. You can navigate back from wherever you go by tapping the back arrow or X in the upper-left corner. To go back to News Feed and your Home page, you can tap the News Feed icon in the bottom area. If you click a link in a post and the bottom area is no longer visible, tap the back arrow to get back to News Feed.

For Android users: The Facebook app for Android has two bars at the top of the screen, not one at the top and one at the bottom.

Search

Although the contents of the top area will change depending on where you are in Facebook, almost every destination has one constant: a search icon, which looks like a magnifying glass. After you tap the search icon, a search box opens that says Search Facebook. (It may also display recent searches you've made.) And the keyboard interface appears, allowing you to type your search.

Depending on where you are on Facebook, the search box may default to a more limited search. For example, if you're viewing groups and tap the search icon, Facebook assumes you're looking for a specific group. The search box will say Search Groups instead of Search Facebook.

Facebook autocompletes when you type, so the items you see below the search box shift as you type. If you're typing a friend's name to get to her timeline, simply tap the person's face when you see it. If none of the autocomplete options fits your search, tap the blue search button on the keyboard interface when you have finished typing. You can then browse all the search results available for your query.

Facebook Search allows you to search not just for people but for places, Pages, posts, and other types of Facebook content. The search results page has an extra bar at the top of the page. Tap any of the filters to see results of only that type.

More

The more icon (three horizontal lines) at the bottom of the screen is where you can see all the links normally found on the left sidebar. Tapping the more icon displays the screen shown in Figure 7-3.

This screen is a catchall, with several section. Some items appear in boxes and others appear in lists. Some sections are hidden behind See More (which you have to scroll down to find). The items you see in any of these sections will vary depending on which parts of Facebook you use and what parts Facebook is promoting. Following is a list of the different items you can find in this space as of this writing:

>> **<*Your Name*>:** This option takes you to your timeline. In general, timelines are organized in the same format as timelines on the regular site, with a few stylistic changes to make sure all the information can fit on the screen.

>> **<*Pages*>:** If you admin any Pages, they appear here. You can tap them to view their timelines. In Figure 7-3, 4 Dummies is a Page that we created.

>> **Suggested News:** Facebook features one article in this box. Tap it to go to the News page, where you can browse news articles based on your interests and news sources you follow.

>> **Marketplace:** Marketplace is a place for people to connect to buy and sell locally.

FIGURE 7-3:
More links.

>> **Saved:** You can save links, articles, and videos for later viewing from your News Feed. When you decide you have time to look through these saved items, tap this option.

>> **News:** The News page features news headlines from official news sources (as opposed to news your friends have posted in News Feed).

>> **Events:** This option lets you view any events you've RSVP'd to as well as browse public events you may want to attend.

>> **Jobs:** Facebook is one of many sites you can check out that show you job listings based on location and industry. You can also post a job that you're looking to fill.

>> **Voting Information Center:** In the lead-up to the 2020 US election, Facebook created this destination for non-partisan information about how to register and vote in your location.

>> **Friends:** The Friends section of the iPhone app resembles your phone's contact list. You can scroll through your friends from A to Z or search for them from the search box at the top of the list. You can also view friend requests and friend suggestions here.

>> **Groups:** View what's been happening in groups you've joined.

>> **Memories:** You need to use Facebook for more than a year for this feature to become interesting, but once you hit that milestone you will be tickled by the memories you can find here. Sometimes the memories of what you posted a year ago are big ones, but often the mundane ones — say, the random photo of your son in his Captain America costume from five years ago — strike a chord and let you take a moment to savor your memories.

>> **Pages:** View a list of the Pages you manage. You can also browse for Pages to follow.

>> **Dating:** Facebook Dating activity is sequestered from the rest of your activity on the site. Tap here to get started filling out your profile and looking for dates.

>> **Gaming:** Browse games to play on Facebook, see updates from the games and gamers you follow, and get started playing games.

>> **Shop:** Check out brands' Pages, view curated items to buy, and scratch that consumerism itch.

>> **Avatars:** Create a cartoon version of yourself that you can use as a profile picture or in gifs and stickers around Facebook. For details, see the "Profile avatars" section.

>> **Blood Donations:** Register as a blood donor and sign up for a local blood drive.

>> **Campus:** College students can find a suite of features here tailored to the campus experience, such as a campus directory, a school-specific News Feed, and dorm chat rooms.

>> **Climate Science Information Center:** Find scientific information about climate change and environmental efforts happening now.

>> **COVID-19 Information Center:** Read vetted information about COVID-19, as well as recommendations from health organizations, information about case rates near you, and any other relevant facts.

>> **Crisis Response:** Whenever a natural (or man-made) disaster occurs, Facebook activates features that enable people to let others know they are safe (this is called *Safety Check*, and we hope you never need to use it), offer or receive help from neighbors or any agencies reacting to the disaster, and fundraise for relief efforts. All these features can be found in the Crisis Response center.

>> **Device Requests:** If you're connecting Facebook to another device such as a smart TV or smart home assistant, you will see requests for that connection here.

>> **Facebook Pay:** You can transfer money between yourself and Facebook friends provided both of you have debit or credit cards or a PayPal account. Go here to add your card info or initiate a transfer.

>> **Find Wifi:** Facebook can search for Wi-Fi networks near you that you can connect to. Using Facebook over Wi-Fi instead of your cellular data network usually makes Facebook load a bit faster and lets you avoid extra data charges.

>> **Fundraisers:** Facebook provides an easy way for you to ask your friends to support a personal or charitable cause, called Fundraisers. Get started here.

>> **Lift Black Voices:** Listen to TED talks from Black leaders and learn about criminal justice (and injustice) in the US.

>> **Live Videos:** Check out videos people are currently broadcasting live.

>> **Mentorship:** Facebook can try to connect you with a mentor (or mentee) for personal or professional growth.

>> **Messenger Kids:** Set up your under-13-year-olds with a version of Facebook Messenger that lets them chat and play games with approved friends and family members.

>> **Most Recent:** Check out the most recent posts from friends and Pages you follow.

>> **Movies:** Check out showtimes and details for recent movie releases.

>> **Nearby Friends:** Opt-in to sharing your location with friends, which increases the chance of a serendipitous run-in when you're out and about.

- >> **Offers:** View coupons and offers that businesses have posted on Facebook.

- >> **Recent Ad Activity:** View ads you've recently interacted with.

- >> **Town Hall:** Virtually every politician — from your local city council member to your senator to the president — has a Facebook Page where they post updates, responses to news, and more. Town Hall makes it easy for you to follow your representatives, send them feedback, and find information related to voting.

- >> **Weather:** Facebook provides access to a weather forecast based on the location of your phone. If you'd like, weather updates can be sent to your News Feed so that you know what to wear in the morning as you blearily check your News Feed.

- >> **Help and Support:** If you have a question that this book can't answer, tap Help and Support to visit Help Center, check out the community help pages, check your Support inbox, report a problem, or read Facebook's terms and policies.

- >> **Settings and Privacy:** Get to your Account settings and privacy shortcuts, learn more about how much time you spend on Facebook, and change the language in which you use Facebook.

- >> **Also from Facebook:** Learn more about Facebook's physical products — Oculus, a VR (virtual reality) headset, and Portal, a smart video-calling device. Unlike the rest of Facebook, these cost money.

- >> **Log Out:** Tap here to log out of Facebook on this device.

REMEMBER

You might see slightly different or additional items in the More section. These may be links to other features or applications in Facebook. Facebook is tailored to you so what you see may be a bit different from what we see.

News Feed

From your Home screen, scroll down News Feed and see what your friends have to say. If new stories appear while you're reading News Feed, a New Stories bubble appears at the top of the screen. You can tap that bubble to jump to the top of News Feed, without manually scrolling.

A News Feed story in the app looks like a News Feed story on the web. As a reminder, every post, regardless of its type, contains the following information:

- >> Name and profile picture of the person (or Page) who posted it.

- >> Time and date of posting. Usually Facebook tries to show you things that happened recently.

- » Location information, if your friend chooses to share it.

- » The content of the post, whether that's a status update, photo, life event, or link to something else on the web.

- » Info about the reactions and comments the post has received so far.

- » Links to Like/React, Comment, and Share (if available).

Tap a photo to see it in the photo viewer. Tap a name to go to that person's timeline or Page. Tap the number of comments to read what's been said about a post. Tapping a link opens the linked article or web page in the Facebook app.

As you scroll down your News Feed, you may notice that videos begin to play as you get to them, but the sound doesn't turn on automatically. You need to tap these videos to play them with sound.

Towards the top of News Feed is the Stories section. Facebook *stories* are a way for people to share snippets of their day without adding them to their timelines. The photos and other content in a person's story disappear after 24 hours. A story can be a fun way to share a ton of photos of a cool hike or some other activity without the content taking over your friends' feeds. We talk more about sharing and viewing stories later in the chapter.

Reacting to Posts

If you like what you see, tap the like icon (thumbs up symbol) at the bottom of the post to let your friend know, hey, I liked that. You can also tap and hold down on the icon (in other words, just put your finger down on it without lifting it back up) to open the full list of Facebook reactions: like (thumbs up), love (heart), care (hugging a heart) haha (laughing emoji), wow (open-mouthed emoji), sad (crying emoji), and mad (red and angry emoji). Figure 7-4 shows the many Facebook reaction icons.

FIGURE 7-4:
Facebook reactions.

No matter where you are on the Facebook app — it doesn't have to be News Feed — you can usually like something in the same way. Tap once to like, and tap and hold to choose another reaction.

Commenting on posts

If you have something to say in response to a friend's post, tap Comment (with a word bubble icon) to open a text box where you can type your comment, as shown in Figure 7-5.

Use your phone's keyboard to type your comment. To add a photo, tap the camera icon next to the comment box. Similarly, tap the GIF icon or smiley icon to add GIFs and stickers, respectively, to your comment. When you're finished with your comment, tap the paper airplane icon to send it, adding it to any other comments that are already attached to the post. Your friend will be notified that you commented on the post.

TIP

Often people are speaking to each other in comments, and to get the attention of certain friends, they will often address a comment to them. As you type a friend's name, Facebook autocompletes the name to *tag* the person in your comment. Tap your friend's name or face when that person appears in the autocomplete list above the comment box. This will officially tag the person in the comment, which creates a notification that your friend will see as soon as he or she logs in.

FIGURE 7-5:
You got something to say?

Post and News Feed options

At the top-right corner of each post is a more icon (three dots). Tap the more icon to open a menu of options for that post, as shown in Figure 7-6.

This menu has the following options:

>> **Save Post:** This option adds the post to your Saved section, so you can easily get back to it later.

- » **Add *<Friend>* to Favorites:** This option prioritizes this friend in your News Feed, so you'll always see anything that person posted at the top of your News Feed in the future.

- » **Hide Post:** If you just don't want to see a post, you can hide it. After you hide a post, it is immediately removed from News Feed. Additionally, News Feed will show you fewer posts like that one. After the post has been hidden, News Feed displays an additional set of options. You can click Undo if you realize you didn't want that post hidden, snooze that person or Page for 30 days, or report the post (more on the last two options later in this list).

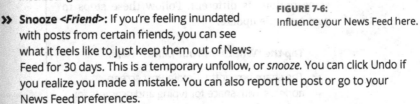

TIP

Liking posts and hiding posts are sort of the yin and yang of adjusting what you see in News Feed. These actions tell Facebook the sorts of posts you like and the sorts of posts you don't. The algorithm adjusts based on what you do so that, over time, you see more and more interesting stories.

FIGURE 7-6:
Influence your News Feed here.

- » **Snooze *<Friend>*:** If you're feeling inundated with posts from certain friends, you can see what it feels like to just keep them out of News Feed for 30 days. This is a temporary unfollow, or *snooze.* You can click Undo if you realize you made a mistake. You can also report the post or go to your News Feed preferences.

- » **Unfollow *<Friend>*:** *Unfollowing* friends means you will no longer see their posts in your News Feed. It is not the same as *unfriending,* which severs the digital link between you and your friends, and prevents them from seeing *your* posts in their News Feed. Unfollowing can be undone from your News Feed preferences.

- » **Find Support or Report Post:** Some posts may contain content that makes you uncomfortable or concerned. Facebook relies on reports from users to take down inappropriate content, though just because you report something doesn't mean it will be removed. If a particular person or Page is the source of multiple items you need to report, you may also want to block or unfollow that person or Page. After you tap this option, the report screen appears, and you can select why you are reporting the post, add requested details, and submit the report to Facebook.

- » **Why Am I Seeing This Post?:** Tap this option to learn more about the factors that contribute to why a particular post has appeared in your News Feed. As an avid reader of this book, you may not find much new information here — News Feed pays attention to the types of posts you like in the past and tries to show you people and posts that you're likely to find interesting.

>> **Turn On Notifications for This Post:** If you're interested in the discussion happening on a certain post, you can choose to turn on notifications for that post. You'll then receive a notification when a new comment or reaction to the post occurs. When you comment on a post, you receive notifications by default when someone comments after you or reacts to your comment.

Posting from the App

Much like on the website, you can post content from the top of your News Feed. Tap What's on Your Mind? to open the mobile share box, shown in Figure 7-7.

The mobile share box has the same options as the share box on the web, although the location of some options is different. Follow these steps to post a simple update:

1. **Tap the "What's on Your Mind?" field.**

   The share box or the Create Post screen appears, with space for typing and prompts for the different information you can add to a post.

2. **Tap where it says "What's on Your Mind?"**

   Your phone's keyboard interface appears. The other options are still available, but you must tap their icons to reveal them.

3. **Type what you want to say.**

4. **Tap Post in the upper-right corner of the screen.**

   Your update is posted to your timeline and shared with friends through News Feed.

FIGURE 7-7:
iPhone posts start here.

A simple status post require just four steps, although you add or adjust a bunch of other options every time you post. For example, in Step 1, you can view the posting options in a menu that overlays the share box (refer to Figure 7-7). To see the entire menu, scroll up. The options in the share box and menu are as follows:

>> **Create Room:** Rooms is Facebook's feature for drop-in video chatting. You can create a virtual room where all your friends can come and hang out.

>> **Photo/Video:** Tap the photo/video icon to add photos or videos from your phone's camera roll to your post. The next section, "Photo Posts," covers a ton of options available with the photos and videos you add to Facebook.

TIP

If you want to add photos you've posted to an existing album, tap +Album, right below your name. You see a list of your existing albums. You can create a new album of the photos you've just selected or add them to an album you've previously created.

>> **Tag Friends:** Tagging people is a way to link them to your post. You might tag friends because they are with you or because something you're posting would be of interest to them. You can add a tag by tapping Tag Friends or by typing the person's name when you're writing your status update. Facebook automatically displays a list of friends when you start typing a friend's name.

>> **Feeling/Activity:** You can add emojis and activity info to your post. Tap to open a list that lets you choose whether you want to share feelings (search through and select smiley face emojis) or activities (search through and add specific things you might be doing, listening to, watching, eating, and so on).

>> **Check-In:** By default, if you've enabled location services when you first added the Facebook app, Facebook will include some general location info, such as your city and state, whenever you post something. A *check-in* is a way for you to select a more specific location such as a restaurant, an attraction, or even your own home.

>> **GIF:** A *GIF* is an quick animated clip that plays on loop, usually without sound. People often use them to express a feeling through the language of pop culture. Tap to search through the GIF library by keyword.

>> **Background Color:** As you type, background color options appear just above the keyboard. Tap one to see what it will look like behind your text.

>> **Live Video:** Broadcast from your phone whatever you are looking at or doing.

>> **Camera:** This option opens your phone's camera so you can take a new photo to share straight to Facebook.

>> **Ask for Recommendations:** Suppose you are out with your phone and are hungry, but you don't know where to eat. You can quickly solicit recommendations from friends. Tap to choose a location, and then describe what you're looking for in the text of your post.

>> **Watch Party:** Create a *watch party* where you and your friends can watch videos together on Facebook at the same time.

>> **Support Nonprofit:** Create a fundraiser to support a non-profit you are partial to. Often people create fundraisers in honor of their birthdays or in response to events happening in the news.

>> **Sell Something:** Tap to create a quick post that includes the name and description of the item you're selling and a price. You'll probably want to add a photo of the item as well.

Even if you bypass this menu in Step 1, you still have access to all the options. After you tap the "What's on Your Mind?" section, the same icons appear above the keyboard, so you can tap the three dots icon to display the menu of post options. You can also select a background color directly from the little preview squares below the text you're typing.

Additionally, you can use the Privacy area below your name to review and change the audience who will be allowed to see your posts. By default, the privacy is set to the same audience as the last post you made.

Photo Posts

Given that so many photos live on phones these days, chances are you'll want to use Facebook to share them. Sharing photos is easy, but a lot of advanced options are available. First, let's go over the basic photo–sharing post:

1. **When creating a post, tap the Photo/Video icon (the green one with a tiny mountain inside a box).**

    This icon is in several places — in the center of the share box at the top of News Feed, in the list that appears after you tap to open the share box, in the icon list when you start typing your post. Regardless of when you tap it, the same thing happens: Your phone's camera roll appears.

2. **Tap the photos or videos you want to select.**

    Facebook highlights selected photos in blue and counts how many photos you've selected, as shown in Figure 7-8.

    TIP

    Tap photos in the order in which you want them to appear.

FIGURE 7-8:
Share photos from your phone's camera roll.

3. **Tap Done in the upper-right corner.**

The share box opens and displays the photos you've added.

4. **Tap Post in the upper-right corner.**

News Feed appears, and you can see your new post added as a new story.

These are the basics. Just a few taps and you've shared a moment of your life with your friends. Now, if you want to explore more advanced topics, including a few photo–editing options, read on:

1. **In the share box, at the top of News Feed, tap Photo.**

Your phone's camera roll appears, which you can use to select photos.

2. **Tap the photos you want to select.**

Facebook highlights selected photos in blue and counts how many photos you have selected (refer to Figure 7-8).

3. **Tap Done in the upper-right corner.**

The share box appears, displaying the photos you've added.

4. **If you have multiple photos, choose a layout from the options at the bottom of the screen.**

By default, photos appear in a modified grid, with some photos larger than other. Your photos can also appear in columns, a banner, or a frame. Tap each option to see how your photos will appear in the post. You can access advanced options also by tapping a photo and then tapping any of the buttons that appear on top of the photo:

- *Edit the photo:* The Edit button in the upper-left corner of the photo opens a number of options covered in Step 6.

- *Make the photo 3D:* This option is available only when you're adding a single photo. Tapping it allows Facebook to re-render the photo so it appears more three dimensional.

- *Access more options:* Tap the three dots icon to display a menu of options. The first two, Remove Photo and Edit Photo, are redundant. The last, Edit Alt Text, allows you to add descriptive text to the photo for people who are visually impaired. This is different than a caption, which might be funny or unrelated to the photo.

- *Remove the photo:* Tap the X in the upper-right corner of the photo to remove that particular photo from your post.

5. **If you're adding multiple photos, add captions to individual photos in the Add a Caption text box below each photo.**

   As soon as you tap the blank box at the bottom of the photo, a keyboard appears so you can type a caption for your photo. When you're finished, tap Done in the upper-right corner.

6. **To edit any photo, tap the Edit button in its upper-left corner.**

   The photo editor appears, with a black background. Along the top of this editor are several icons for editing, as shown in Figures 7-9, left:

- *Tag:* Tags are ways to mark who is in a photo. Tap a face, and the face becomes highlighted and a text box appears so you can type the person's name. Facebook autocompletes as you type; tap the correct name when you see it appear. Figure 7-9, left, shows what tags looks like in a photo.

- *Crop:* Like any cropping interface, this one allows you to change the edges and orientation of your photo. Use your fingers to move the proposed border of the photo (see Figure 7-9, middle). You can also rotate the photo (no sideways photos for you!) and automatically resize it to a square instead of a rectangle.

FIGURE 7-9: Tag, crop, and add stickers to your photos.

- *Stickers/GIFs/emojis:* Tapping this icon opens a list with a wide range of additions to your photo: location tags, timestamps, tags of people, music clips, photos from your camera roll, feelings, seasonal stickers, a full list of GIFs, and all the usual emojis. Think of these options as a decoration that is placed on top of your photo. You can add lots of different stickers in a variety of combinations by returning to this list over and over again (see Figure 7-9, right). After you have tapped the item you want to add, you return to your photo in the photo editor, with the item now stuck in the center of your photo. Use your finger to drag the item. You can also tap to change the style of many text-based stickers. If you decide you no longer want one of these decorations, drag it to the trash icon in the bottom of the photo editor.

- *Text:* The Text option displays text on top of the photo, as shown in Figure 7-10, left. (Captions, described in Step 5, are displayed below your photo.) Tap the text icon and type the text you want to add. Tap the font menu to switch the font — Facebook offers a limited suite of fonts with descriptors such as Pop and Fancy. Tap the color wheel to change the color of the text. Tap the bars icon, which may be familiar to you from your word-processing program, to change whether the text is right or left justified. Tap the little letter in a box to change the effects around the text (depending on the font you chose).

  The text immediately reflects any changes you make to its appearance, so you can easily try out a few options. When you like the text, tap Done. You can then drag the text to a different location on your photo (or drag it to the trash icon to delete it).

- *Doodle:* You can doodle a little drawing on the top of any photo, as shown in Figure 7-10, middle. When you tap the doodle icon, a line thickness guide appears on the right side of the screen. Drag your finger up and down on that guide to select the line thickness. Tap the paintbrush icons at the bottom of the screen to change the quality of your lines (opaque, spray-paint, or eraser). Tap the palette icon to select a color for your doodle. You can change the line color at any point in the doodling process. Draw your doodle, using your finger, on top of your photo. Use the Undo button at the top of the screen to undo your lines.

- *Effects:* Effects are ways to change the overall color or tone of the photo. For example, you can convert a photo to black-and-white, or make it look as though it's a charcoal drawing, as we did in Figure 7-10, right.

- *Lighting:* Lighting allows you to change the brightness level on your photo, using a slider to make the photo brighter or darker.

- *Save:* If you've added any of the previous options to your photo, you may want to save a version of it to your phone. That way, the edited version doesn't live only on Facebook.

7. **When you have finished editing your photo, tap Next in the lower-right corner.**

8. **Repeat Steps 6 and 7 as needed for each photo you're posting.**

9. **When you have finished with your edits, tap Done in the upper-right corner.**

   You return to the mobile share box, just where you were after Step 3.

10. **(Optional) To add a photo to an existing album or create a new album:**

    a. *Tap* the +Album button. It's located at the top of the post, right under your name. You see a list of your previously created albums.

    b. *Tap the album you'd like to add to, or tap to create a new album.*

    c. *If you're creating an album, name it.* You return to the mobile share box, and the name of the album is displayed at the top of the screen.

11. **Double-check who can see your photos at the top of the post.**

    A grey box at the top of the share box is prefilled with your most recent privacy choice. You can change the privacy for this post by tapping and selecting a new privacy option from the menu that opens.

12. **Tap Post in the upper-right corner.**

    You return to News Feed, where you can see the News Feed story that has been created about your new photos.

Taking photos and creating videos to share

While Facebook defaults to displaying your phone's camera roll when you go to share a photo, often people skip the middleman and take photos straight from Facebook to share. *Selfies,* or pictures you take of yourself with your phone's camera, are one of the most basic ways of documenting your life. Facebook provides a lot of options to add pizzazz to photos and selfies when you take them through the Facebook app.

To understand the various options available to you, let's take a closer look at the phone screen when you're using the camera through Facebook. You can get here by tapping Photo in the share box at the top of News Feed. Your phone's camera roll appears. Tap the grey camera icon in the upper-right corner of the screen to open your phone's camera through Facebook.

Parts of this screen shown in Figure 7-11, middle, should look familiar if you've used your phone's camera. You take a picture using the big button in the center. On either side of the button are the flash and flip camera icons, which allow you to use the flash and to flip whether you are recording video from the rear camera or the front camera (as you would for a selfie). The rest of the options might be new, but we hope they won't seem overwhelming by the end of this section.

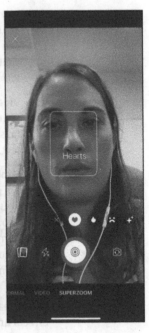

FIGURE 7-11:
Start a live video, take a selfie, or use superzoom.

At the bottom of the screen are several options for the type of photo or video you're taking. Swipe to see all the options:

>> **Live:** Live videos are what they sound like: a way of broadcasting yourself (or what you're looking at) live. You can add effects to your video, conference in a friend, or start a live shopping or donation link (see Figure 7-11, left). Some people wonder what they should be sharing. The answer is, whatever you want. Maybe you want to broadcast the concert you're watching, or give a glimpse of the waterfall you just hiked to, or share your toddler's tantrum. Celebrities and other Pages often use live video to give a behind-the-scenes peek to their fans or followers. Plenty of news stories have been based on live broadcasts. Just because you aren't a celebrity doesn't mean that you can't share a behind-the-scenes peek of your own life.

>> **Boomerang:** A *boomerang* is a type of video you record that mimics a GIF; it's a short clip that repeats itself. When you take a boomerang, make sure you're aiming the camera at something that's moving. Boomerangs are only 4 seconds long, which makes them ideal for capturing something quick and flashy: the dog jumping to catch a treat, your friend attempting a trendy dance move, the exact moment you blew out the birthday candles. You can add effects, stickers, doodles, and text to your boomerang just as you would a normal photo.

>> **Normal:** Normal is the default when you open your phone's camera. (See Figure 7-11, middle.) When you flip the camera around for a selfie, you may see additional options for color washes (warm, cool, glow) above the big center (take picture) button. These optional lens filters can make your face look its best.

>> **Video:** This option is the equivalent of Normal but for video. Tap the record icon in the center of the screen to start recording, and tap it again when have finished. After you record your video, you can edit it as well as add stickers, text, doodles, and effects.

>> **Superzoom:** This option, another specialized video clip, zooms in where you are pointing the camera. You can add special effects to really tell a story. See Figure 7-12, right. For example, when a 5-year-old decides to cry extensively about a sandwich being cut into triangles instead of squares, adding the Bummer filter, which turns the screen blue and shows leaves falling to the ground as the camera zooms in on the tears, drives home just how tragic this moment was.

If you read the last section about photo posts, you know that Facebook has many effects and filters that can be applied after you've taken your photo. In addition, when you're taking selfies, you can play around with effects and masks. The smiley face icon that appears to the right of the flip camera icon indicates that you can add these specialized masks on boomerangs, videos, and regular photos. While filters can be applied to the content of any photo, many of the masks require the camera to recognize a face. A few examples of these mask effects are shown in Figure 7-12. Tap the smiley face icon to open a list of options at the bottom of the screen. Tap the one you want to try to see if you like it.

TIP

If you want to impress the small children in your life, let them see themselves with various masks. As Carolyn's kids like to say: "Do effects! Do effects!" Don't share any photos of them without their parents' permission, though!

After you've settled on the type of photo or video you want to take, tap the big center button to take your photo or record your video. The photo or video is then displayed on the screen. If you want to retake it, tap the back arrow in the upper-left corner. Everything else is the same as the a regular photo post (crop, stickers, text, doodle, effects). An additional icon allows you to save the photo to your phone's camera roll.

FIGURE 7-12:
Funny faces.

If you're happy with your photo, you can share it by tapping the Next button in the lower-right corner. This brings you back to the Create Post screen, which should look familiar. (If it doesn't, head back up to the "Photo Posts" section.) Then tap the blue Post button in the upper-right corner.

Creating Facebook stories

Facebook *stories* are a way to share several tidbits of your day with your Facebook friends without creating a permanent Facebook post that will live in perpetuity on your timeline. You can add many things to your story at once or add things to it as your day progresses. It's a fun way to document whatever's happening on a given day without worrying about the quality of the photo you're sharing or feeling the need to provide the perfect clever caption.

In Chapter 4, we go over how to add stories from your computer, but stories are meant to be shared from your phone. You can add lots of options, and share in several ways. Many options will be familiar if you've been reading this chapter in order.

First, let's go through the process of creating a basic story: a quick photo of something you encounter during your day. To create a story, follow these steps:

1. **From your Home page, tap Create a Story, the leftmost item in the Stories section.**

   The Create Story screen appears, as shown in Figure 7-13. The top of the screen shows different story types you can create. (Skip down a few paragraphs for a description of story types.) The bottom two-thirds of the screen show your phone's camera roll. You can also tap the camera icon in the bottom center to open your phone's camera.

2. **Tap the photo you want to add to your story.**

   The photo is enlarged on an edit photo screen, where you can add stickers, text, doodles, effects, tags, and zoom effects. You can also change who can see the story or save the story to your phone's camera roll.

3. **Tap the blue Share to Story button in the lower-right corner.**

   Your story is published and you — and your friends — can now see it in the Stories section at the top of your News Feed.

FIGURE 7-13:
Start your story here.

Adding stories can just as easy as creating a normal post, and there isn't any need to make it absurdly complex or decorative. At the same time, you might want to make your story absurdly complex and decorative. First, there are many types of stories you can create. You can select any of these from the top of the Create Story screen (refer to Figure 7-13). To see all the options, scroll across the screen. When you've figured out the type of story you want, just tap it:

» **Text:** A text story is just what it sounds like. Choose the color background by tapping the small colorful circle, and then type the message that you want to appear in your story. Again, there are no rules on what to share here — say whatever you want.

» **Music:** Do you want to tell people that you finally dropped off the kids for the first day of school and have the house to yourself, or do you want to blast Lizzo's "Good as Hell" and have people feel what you feel? Sometimes music can say a lot on your behalf. With a music story, you choose a song as your story's foundation. Facebook has a wide library of songs you can choose from, including most popular music. You can browse for songs or use the search function to look for the song you want to choose. If the song has the word *lyrics* next to it, you can display the lyrics in the story as well as play the song.

After you've selected your music by tapping it, you see a preview of your story. Tap the song lyrics or album art to change how the song is displayed — lyrics or not, large album art or small, even which clip of the song to play. Like any other story, you can edit the color of the background, as well as add effects, stickers, text, and doodles.

>> **Boomerang:** A *boomerang* is a video-photo hybrid that you can take from your phone's camera. It creates a 4-second clip that automatically replays as your friends view your stories. When you take a boomerang, make sure you point the camera at something that is moving. You can then add stickers, text, doodles, effects, and tags to the boomerang.

>> **Mood:** Mood stories use GIFs as the base. While you can always add GIFs to other stories, mood stories start with the GIF and build out from there. When you tap the Mood option, you will see a list of mood GIFs, sorted by theme. Browse through this list or search for a word.

After you've chosen your mood, you see the story preview with that GIF front and center. Tap the GIF to change its appearance. You can also change the background color behind the GIF (even selecting a photo from your phone's camera roll). Tap the icons to add stickers, text, doodles, and effects.

>> **Selfie:** If you start a selfie story, Facebook automatically opens your camera so that you can take a photo of yourself, similar to Figure 7-11, middle. After you take your photo, you can add stickers, text, doodles, and effects if you want.

>> **Poll:** Poll stories allow you to ask a question that people viewing your story can respond to easily. By default, when you start a poll story, the options are yes and no. To edit the text to ask any question that has two possible answers, tap Yes or No and then change the text inside the poll button. You can change the background of the poll, even using a photo from your phone's camera roll as the background to, say, take a poll about something happening in the photo.

The other ways to make your story pop are the same as those in the photo post toolkit: Add decorations (such as sticker)s, text, doodles, and effects. You can add as many photos (or other types of stories) to your story. You can mix these many types of story posts together, or focus on just one type. Also, remember that all this will be gone tomorrow. As you start using stories, feel free to experiment with the different types of posts and the types of things you like sharing in this format. If it doesn't work for you, no problem — it will disappear from Facebook in 24 hours.

Although stories will disappear from Facebook in 24 hours, they won't necessarily disappear from your friends' memories. Also, if someone who can see your story takes a screenshot of something you said in a story, it can have a long life beyond its time on Facebook. Stories aren't a good place to post a screed against someone you dislike, thinking that it won't get back to that person.

Friends who are using Facebook during the days you share stories can view your stories and react to them. Instead of leaving comments, friends can quickly send you a message from your story to let you know what they think about it. You'll receive notifications about these responses alongside the rest of your notifications.

After your story has expired, you'll still be able to find a record of it, who saw it, and people's reactions to it in the Archive section of your timeline.

Viewing and interacting with stories

The Stories section contains a series of vertical boxes showing previews of the stories you can view, as shown in Figure 7-14. If your friends haven't posted Stories in the last 24 hours, this section may be collapsed.

Tapping a story opens it, filling your screen, as shown in Figure 7-15. Stories play like a slideshow, with each "slide" getting a fixed amount of time on the screen before automatically moving to the next post. If you think the story is moving too fast, tap and hold down on the screen to pause the story's progress. Release your finger to let the story continue playing. If you feel like the story is moving too slowly, a quick tap on the right side of the screen advances the story to the next post. A quick tap on the left brings you back to the previous slide.

— Stories

FIGURE 7-14:
Stories can be found at the top of your News Feed.

At the top of the story, a series of thin white lines indicates how many posts the story contains. The lines change from thin to thick as the story progresses through its posts. Use the comment box and reaction emojis at the bottom of the story to react to what you're seeing. Comments on stories generate a private message to the person who created the story, unlike comments in News Feed, which everyone can see. If you get bored looking at stories, tap the X at the upper-right corner of the screen.

Checking Out Timelines

Like News Feed, timeline in the Facebook app should look familiar, if a bit narrower than you're used to. Figure 7-16 shows a friend's mobile timeline. A cover photo sits across the top of the screen, with a profile picture in the center. You can see some icons to interact and some basic biographical info.

On your own timeline, you'll see options below your profile picture to Add Story or see more options (the three dots icon). Tapping the three dots icon displays a menu of options where you can edit your profile, see your story archive, view your profile as if you are someone else, review your Activity Log and timeline, adjust your privacy settings, search your profile for a specific post, and adjust your memorialization settings.

Below the Add Story button is some of your About Me info. Scroll down and you'll pass by featured photos, a preview of your friends, and the mobile share box, where you can create posts. Next are your posts. Most recent posts appear first, and you can keep scrolling back to the very first post.

FIGURE 7-15:
Watch a friend's story unfold.

FIGURE 7-16:
A friend's timeline in the Facebook app.

When you view posts, you can tap the three dots icon in the upper-right corner to see options related to that post (see Figure 7-17). You see different options depending on the type of post, but in general, this is where you can hide a post from your timeline, edit the content of a post, change the privacy settings on a post, and more.

Profile videos

One mobile-only feature you can add from the Facebook app is a profile video. *Profile videos* are like a profile picture but in motion. Profile videos are up to 7 seconds long and play on a loop when people are looking at your timeline from their phone. A still image from the video will then appear as your profile picture in thumbnails around the site. Follow these steps to record a profile video:

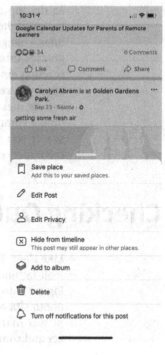

FIGURE 7-17:
Post options.

1. **From your timeline, tap the camera icon at the bottom of your profile picture.**

    A menu appears for changing your profile picture and profile video.

2. **Tap Select Profile Picture or Video.**

    Your phone's camera roll appears.

3. **Choose a video or record one:**

    - *To choose an existing video:* Simply tap it.

    - *To record a video:* Tap the camera icon in the upper-right corner to open your phone's camera interface. Tap the video camera icon in the bottom right to switch the camera from photo to video. (The center icon turns from blue to red.) Tap the red circle to begin recording, and tap it again to stop recording.

    Remember, you have a maximum of 7 seconds to work with. Do whatever comes naturally.

    The Preview Profile Video screen appears, with a preview of your new profile video.

4. **Tap Edit below the video preview to edit the video.**

5. **Edit your video as needed.**

   You can trim the video, frame by frame, crop the video so your face is perfectly centered, turn the sound on or off, and choose the cover, or still frame, from the video, which will be used as your profile picture in comments and elsewhere around the site.

6. **When you're ready, tap Done in the upper-right corner.**

   You return to the Preview Profile Video screen. You can repeat Steps 4 and 5, or tap to add a frame to your video. You can also choose to make your profile picture temporary. (If these features aren't familiar, flip back to Chapter 5, where we detailed profile picture features.)

7. **When you are sure you're ready, tap Use in the upper-right corner.**

   You return to your timeline, where you can see your profile video in its natural habitat.

REMEMBER

Profile videos, just like your profile picture and cover photo, are visible to anyone who visits your timeline. Thumbnails from your profile video appear in News Feed and anywhere else on Facebook where you might see a profile picture.

Profile avatars

Another feature you can access only on your mobile phone is a profile avatar. An *avatar* is a cartoon you can create and modify on Facebook to look like you. Carolyn's avatar in shown in Figure 7-18. You can use your avatar in comments and messages with friends.

To create an avatar, follow these steps:

1. **Tap the camera icon at the bottom right of your profile picture.**

   The profile picture menu appears with the following options: Select Profile Picture or Video, Add Frame, and Create Avatar Profile Picture.

2. **Tap Create Avatar Profile Picture.**

   You see the Create Avatars screen or your existing avatar if you already created one. You can start over by tapping the edit icon and then the trash icon, which deletes your existing avatar.

FIGURE 7-18:
Say hi to Avatar
Carolyn.

3. **Create your avatar by following the prompts on the screen.**

Choose the skin tone, hair style and color, face shape, complexion, age, eye shape and color, eyebrows, accessories such as eyeglasses, bindis, or hats, nose shape, mouth shape and color, facial hair, body shape, clothing, head-wear, and earrings. Scroll through the options and tap on the ones you want to try. You can tap the mirror icon in the upper right to look at yourself through your phone's camera while you make these choices.

4. **When you're finished, tap the blue Done button in the upper-right corner.**

You see a preview of your avatar. You can share a post about your new avatar or turn it into your profile picture to show it off.

You can use your avatar as a type of sticker in messages and comments. Tap the sticker icon (it may now look like your avatar's face!) to see a suite of stickers containing your avatar (refer to Figure 7-18). Tap the one you want.

Using Groups

You can use Facebook groups to communicate with a group of friends about anything in the world. (Groups are covered in detail in Chapter 10.) Do the following to access any group you belong to through the Facebook app:

1. **From the Facebook app, tap the Groups icon (three people) at the bottom of the screen.**

   The Groups screen appears. At the top are buttons you can tap to view all your groups, discover new groups, or create a group, followed by thumbnails representing the groups you belong to. Next is a feed of posts to the groups you belong to, with the most recent on top. Scroll down to read more posts.

2. **Tap on the group you want to go to.**

   You see the group's mobile Home page.

Unsurprisingly, your group's Home page (see Figure 7-19) looks a lot like its Home page on the website. The cover photo is on top, followed by more info about the group, a blue Invite button if you want to add members, and buttons to start a room or a chat, start a watch party, and share photos with the group. Scroll down to get to the share box and the group feed of posts that members have shared recently.

Tap the share box (the Write Something box) to post something to the group. You can post links, articles, text, or photos, and they will be available to all members of that group.

TIP

Just as you can on the website, you can search in a group's posts for information. Tap the search icon (at the top of the screen), and enter a search term (or terms) in the search box that appears. The search results you see come only from the group you are currently viewing.

FIGURE 7-19:
A group on your phone.

Viewing an event

You probably won't use Facebook events much when you're out and about. But when you want to view an event on the Facebook app, it's because you *really* need to see that event. You need the address, or the start time, or a number to call, and the only place you know for sure you can find it is from that event's Page. To get to an event, follow these steps:

1. **From the Facebook app, tap the more icon (three bars) at the bottom of the screen.**

   You see a list of apps and favorites that you might want to visit.

2. **Tap Events.**

   The Events screen appears, displaying upcoming events you've been invited to, upcoming public events that are popular in your area, and upcoming birthdays.

3. **Tap the event you're looking for.**

   You see that event's Page, where you can see information about the event and its guests, as well as publish a post to the event.

Facebook Messenger

The Facebook app is designed to let you do on your phone everything you can do on the computer-based Facebook. One big part of using Facebook is communicating with friends. Communication can happen not only through comments and posts but also by using Facebook messages. (For the details of using Facebook messages, see Chapter 9.) Because getting in touch directly with a friend is sometimes more important when you're out and about than, say, checking his latest photos, Facebook has created a separate app for messages. If you use the Facebook app and try to send a message, you'll be asked to add the Facebook Messenger app as well.

Messenger handles all the chats and messages between you and your Facebook friends when you're on your phone. As a stand-alone app, you can easily get to it and use it to communicate on the go, without getting distracted by the rest of Facebook. Carolyn and her husband use Messenger as their chief means of communication during the day while they're at work. It's just as immediate as texting but doesn't incur additional texting charges on their phones, and it's easy to switch from a chat on the computer screen to a chat in Messenger.

Messenger has the same functionality as sending Facebook messages: You can message one person or multiple people, and send photos, videos, stickers, and GIFs. In addition, conversations are grouped in one place.

The Messenger app described and shown in this section is used on an iPhone. There may be a few slight differences between it and the Android version.

REMEMBER

Navigating Messenger

When you open the Messenger app, you see your recent messages, as shown in Figure 7-20.

The bulk of the page is taken up by your messages, organized by conversation, with the most recent one at the top. Each conversation shows the profile picture of the friend (or friends) you're conversing with. Any conversations that have unread messages appear in bold. Each conversation has a time or date next to it reflecting the last time anyone on that thread sent a message. Scroll down to older conversations. Tap any conversation to view its contents or add to the conversation.

Above the conversations are little bubbles showing friends who are currently online.

At the top of the screen is a search bar that lets you search for people and groups you might want to message. Tap the paper and pencil icon in the upper-right corner to open a new message. Tap your own face in the upper-left corner to open your settings.

FIGURE 7-20:
The Messenger inbox.

Notable settings you can access here include the following:

- » **Dark Mode:** To reduce battery usage and eyestrain, you can turn Messenger to Dark mode, where the background will appear black instead of white.

- » **Active Status:** You can choose whether or not friends can see that you are active on Facebook Messenger. By default, friends see a green dot next to your name if you're currently using Messenger.

- » **Message Requests:** If someone you aren't friends with yet sends you a message, the request goes to a separate inbox. You can get to this type of message from this settings menu.

- » **Notifications & Sounds:** You can turn on a do not disturb setting so you won't be bothered by notifications from incoming Facebook messages. You can also choose whether or not you want to see previews of the content of your messages, and whether or not you get sounds and vibrations while you are using the Messenger app.

When you use Facebook messages, new messages simply get added to any previously sent messages, sort of like text messages on your phone. They get sorted only by who is talking to one another, not by subject matter or date.

At the bottom of the screen are two icons:

>> **Chats:** Tap this icon at any point to return to your recent messages. The number in the bubble represents the number of unread messages.

>> **People:** Tap this icon to see a list of your friends who also use Messenger. You can also view your friends' stories here.

Viewing and sending messages

Figure 7-21 shows a message thread, or conversation, in the Messenger app. It looks a lot like a text message thread, with each message in a speech bubble. Your messages appear on the right, in blue bubbles, and messages from friends appear on the left, in grey bubbles. The most recent message is at the bottom of the screen, and you can scroll up to view older messages. Your entire history of messaging with this friend can be found here. It doesn't matter whether you were messaging from a phone or a computer; you can go back to the beginning of it all.

At the top of the screen, tap the back arrow to go back to your recent messages, or tap the phone icon or video player icon to start a phone or video chat, respectively.

At the bottom of the screen is a space for typing your message. Use the keyboard to enter your message and then tap the send icon (a paper plane) in the lower right of the text you've entered. Tap the arrow to the left of the text box to see the options for content you can add to your message.

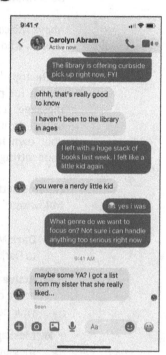

FIGURE 7-21:
A message thread between friends.

Next to the text box you may notice a series of icons that show off the variety of content you can append to a message:

>> **More icon:** Tap the more icon to view more options that can integrate with Messenger:

- *Create Room:* A *room* is a virtual hangout space to which you can add multiple friends. You can create a room with the person you're currently chatting with, and then have other friends join and leave as needed.

- *GIF:* A *GIF* is a file format that has come to describe short (often only a second or two) looped videos. GIFs capture emotions or sentiments that can be hard to express in words (or easy to express in words, but more fun to describe in motion). For example, you might tell someone that something she said is exciting, or you might send a GIF of a unicorn shooting a rainbow out of its horn. Tap the GIF icon to browse, and then tap the GIF you want to send.

- *Payments:* You can request and send money through Facebook. If this is your first time using payments you need to add a bank or credit card to your account. Facebook facilitates the transfer between you and your friend.

- *Share Location:* This option is the solution to the classic "I'm on your street, where are you?" problem. You can share your location with your friend for an hour, making it easy for you two to find each other.

>> **Take Picture (camera icon):** Tap the camera icon to open your phone's camera and take a photo or video to send to your friend. You can use any Messenger filter (covered in the upcoming "Video calls" section) to add a little flair to your photos.

>> **Send Picture (photos icon):** Tap the photos icon to attach photos from your camera roll. Then browse through your camera roll and tap the photo you want to send. Then tap Send (in the upper-right corner). You can edit a photo (by adding a text overlay or a doodle) before sending it.

>> **Voice Message (microphone icon):** Tap this icon to record an audio message to send to your friend. Tap and hold down on the microphone icon to start recording, then let go when you're finished recording. Keep in mind that you won't be given a chance to review your recording; it just gets sent as soon as you're done.

>> **Stickers (smiley icon):** Tap the smiley face icon to add stickers to your message (you can also access GIFs, which are also in the More menu). You have thousands of sticker options. Simply tap to select one. If you want to search more widely, tap the magnifying glass icon at the bottom of the screen (after you tap the smiley icon) to browse stickers by category.

>> **Emoji icon:** Tap this to send your friend the emoji visible to the right of the text box. Often, this is the Facebook like icon by default, which appears as a thumbs up in the message thread.

Video calls

One of the best parts of using Messenger is the ability to easily start voice and video calls with your friends. You don't need to be using the same type of phone — and one of you can be on a computer. To start a video call, open a conversation with the friend you want to call, and tap the video camera icon in the upper-right corner of the screen.

While you're talking to your friend, you have the option to add photo filters to your video. Tap the photo filters icon (a magic wand over a face) to open the photo filters, which you can scroll through from side to side to browse. Facebook groups effects into three categories:

» **Effects:** Effects are generally masks that are applied to your face. An effect can make your eyes appear unnaturally large, or make it look like a cat is on your head. Some masks are seasonal or tied into a movie promotion. They are a good bit of fun.

» **Backgrounds:** Effects change your face, but backgrounds change what is behind your face. You might change the background to an underwater scene, in which case the lighting also changes so your face looks like it's underwater. Facebook is always adding new backgrounds, often depending on the season.

» **Lighting:** Lighting changes make it seem as though you applied a filter to your camera's lens. These can change the temperature of the photo by changing the color balance of the image.

You can take a photo of a moment in the conversation by tapping the white circular button at the bottom of the screen. The screenshot is automatically saved to your phone's camera roll.

You can video chat with more than one friend at a time, provided they all have access to Messenger on their phone or a webcam on their computers.

The Facebook Family of Mobile Apps

In addition to the Facebook app and the Messenger app, Facebook has other stand-alone apps to enhance specific portions of the Facebook experience.

Instagram

Instagram, which is owned by Facebook, is a social media app used for sharing photos with friends and followers. If mobile photography is your thing, Instagram might be just what you're looking for. To share photos, some people use only Facebook, others use only Instagram, and still others use both. You can sync your accounts so that whatever you share on Instagram goes straight into Facebook's News Feed. You can also connect your accounts such that messages sent to you through Instagram appear in the Facebook Messenger app.

WhatsApp

WhatsApp is a messaging app owned by Facebook. It has many of the same features as Facebook Messenger, such as group chatting and video chatting. You don't need a Facebook account to use it, however, which can make it convenient for keeping up with people who don't (gasp) have a Facebook account. Additionally, WhatsApp encrypts all its messages, and this security feature prevents anyone except the designated recipient from accessing your messages.

Messenger Kids

Messenger Kids is a special app for children who are not old enough to have a Facebook account but are old enough to want to keep up with friends. Parents can create a profile for their kids and then approve other friends for them to message and video chat with. Kids can play games together and generally be kids together, remotely.

Facebook Local

Facebook Local is an app that promotes events and businesses close to you. It recommends activities, events, restaurants, and more based on friends' recommendations. If you use Facebook a lot to find public events you want to attend, this is the app for you.

Facebook on Your Mobile Browser

Viewing a web page from your phone can be difficult because information that is normally spread across the width of a monitor must be packed into a tiny column on your phone. Facebook is no exception, which is why the first tip in this section is: Never go to www.facebook.com on your mobile phone. You'll regret it.

But fear not, you still have a way to carry almost all the joys of Facebook in your purse or pocket. On your mobile phone, open your browser application and navigate to m.facebook.com — a site Facebook designed to work on a teeny-tiny screen.

The first time you arrive at m.facebook.com, you're asked to log in. After that, you never (or rarely) have to reenter your login info unless you explicitly log out from your session, so be sure you trust anyone to whom you lend your phone.

WARNING

If you plan to use the Facebook Mobile site frequently, we recommend that you have an unlimited data plan that allows you to spend as much time on the mobile web as you like for a fixed rate. The Facebook Mobile site is nearly as comprehensive and rich as the computer version. You can spend hours there — and spend your life savings, too, if you're paying per minute.

Mobile Home

After you log in to Facebook Mobile, you see the mobile version of the Home page, as shown in Figure 7-22.

TIP

To follow along with this section, you can navigate to m.facebook.com on your web browser. Just imagine what you see on about one-tenth of the screen.

In this section, we detail what you see on the Mobile Home page; we cover the other pages in the following sections.

From m.facebook.com you'll see the following items on your Mobile Home page:

FIGURE 7-22:
Facebook Mobile Home.

» **Blue bar on top of the page:**

The blue bar contains many of the links you need most often for navigating Facebook:

- *News Feed:* Tap the News Feed icon (a little newspaper) in the blue bar on top at any time to see or refresh your News Feed. News Feed itself takes up the bulk of your Mobile Home page.

- *Friend requests, messages, and notifications:* These three are used throughout the Facebook experiences. New or unread actions are marked by a red flag in each category, just like when you log in to your Facebook account from a computer.

- *Search:* Tap the magnifying glass icon to open a search page where you can search for friends, Pages, posts, and trending topics.

- *More:* Tap the more icon (three horizontal lines) to open a menu with links to the options usually found in your left sidebar: your profile, events, groups, favorites, apps, and so on.

>> **Share box:** Facebook makes it easy to update your status, share photos, and check in so you can spread the news the moment you're doing something you want people to know about. The share box on your phone has the same privacy options; just tap or select the people icon on the left side of the share box to change who can see your post.

>> **News Feed:** News Feed takes up most of the space on the Mobile Home page. Scroll down to see more stories from your friends. Just like on the regular site, you can like, comment on, and share your friends' posts.

Mobile timelines

Timelines on Mobile Facebook resemble timelines on the regular site, except you have only one column instead of two. As you scroll down a mobile timeline, here's what you'll see:

>> **Profile picture and cover photo:** The profile picture and cover photo are smaller versions than those on the regular site.

>> **Action links:** If you're looking at a friend's timeline, you see blue links indicating that you're friends and are following that person. If you aren't yet friends with that person, you see links to Add Friend or Message the person.

>> **Photos and friends:** Before you start seeing your friend's posts on her timeline, you can see thumbnails of recent photos she has added or been tagged in, as well as a preview of her friends list.

>> **Share box:** Use the mobile share box to leave a post on a friend's timeline or share a photo there. Remember, things you post on a timeline are visible to your friend's friends.

>> **Timeline:** As on the regular site, the star of the show is your friend's timeline, where you can see her most recent post, whether that's a status update or a photo. You can also see how many people liked or commented on her posts, and you can add to these counts yourself by doing the same.

Mobile inbox

The *mobile inbox* functions the same as the inbox on the regular site, but in a more compact view. In the mobile inbox, your most recent messages appear first. You can see the name and profile picture of the people on a given thread, as well as a preview of the most recent message sent.

Tap a message preview to open it. When you open a mobile thread, as with regular Facebook, the newest message is at the bottom with the Reply box below it. Tap the action icon (an envelope with an arrow) in the upper-right corner to choose from the following options: Mark Read/Unread, Delete, Delete Selected, Move to Done, Ignore Messages, Block Messages, and Something's Wrong. The Mark as Unread option is particularly handy because often you read a message on your mobile phone but don't have the time or energy to type a response right then. Marking it as unread reminds you to respond when you return to your computer.

Facebook Texts

In much of this chapter, we assume that you have a smartphone or at least a phone with some capacity to use a browser. If you don't, you can still use Facebook via text message. To get started with Facebook texts, enter and confirm your phone number in the Settings page from your web browser:

1. **In the upper-right corner of the big blue top bar, choose the account menu (down arrow) and then choose Settings and Privacy.**

The Settings and Privacy menu appears.

2. **Choose Settings, and then click the Mobile tab on the left side of the page.**

The Mobile Settings page appears.

3. **Click the green Add a Phone button.**

You may be prompted to reenter your Facebook password. When that's all squared away, the Confirm Your Number dialog box appears.

4. **Choose your country and mobile carrier, and then click Next.**

TIP

If your carrier isn't listed, you may be out of luck using Facebook texts from your mobile phone.

A space appears for a code to be entered.

5. **Pick up your phone and text the letter F to 32665 (FBOOK).**

FBOOK texts you back a confirmation code to enter from your computer. This step can take a few minutes, so be patient.

6. **Enter your confirmation code in the empty text box.**

7. **If you see the Share My Phone Number with My Friends check box, choose whether you want your phone number added to your timeline.**

It can be useful when friends share their mobile numbers on Facebook because it allows people to use Facebook as a virtual phone book. If you're not comfortable with that, simply deselect the check box.

8. **If you see the Allow Friends to Text Me from Facebook check box, select whether you want friends to be able to text you from Facebook.**

If you don't want people to be able to text you through Facebook, simply deselect the check box.

9. **Click Next.**

This confirms your phone.

After your phone is confirmed, Facebook texts are the most basic way to use Facebook on your phone. Facebook texts are mostly a way to stay updated on actions others are taking on Facebook; you can't update your profile and take your own actions via text.

Here are the various actions you can take on Facebook via text message. All messages get sent to 32665 (FBOOK):

>> **Get a one-time password for accessing Facebook:** Text the letters OTP. One-time passwords allow you to access Facebook from a new computer without accidentally letting your password be saved and giving someone else access to your account. Just remember to log out when you're done!

>> **Stop getting texts.** Text the word **stop.**

>> **Get help:** Text the word **help.**

>> **Start getting texts.** Text the word **start.**

Mobile settings

After you start using Facebook texts, you might want to adjust certain settings to better suit your texting lifestyle. From the Settings page, select Mobile on the left side of the page. You have access to the following settings:

>> **Text Messaging:** Decide which phone number you want your texts to be sent to. You need to change this setting only if more than one mobile phone number is listed for your account.

>> **Daily Text Limit:** The Daily Text Limit allows you to modify the number of text messages you receive per day. Remember to click Save Changes after updating this setting.

WARNING

If you have a mobile plan for which you're charged per text message (and you're exceedingly popular), use the setting that limits the number of messages Facebook sends you per day. Otherwise, you may have to shell out some big bucks in text message fees.

>> **Time of Day:** You can choose to get texts from Facebook at only certain times of the day, so that your phone doesn't bother you when you're sleeping (or while you're at work, perhaps).

Mobile notifications

The Facebook text experience is mostly about being notified about actions friends take on the site, so it's important to understand where to go to change which of these notifications are texted to you. To get started, head to the Notifications tab of the Settings page, which displays all the types of notifications that Facebook generates for you. Click a type to view its settings. For each type, you can decide if you want to receive those notifications and how you want to receive them: on the site, by email, or by text. For each type you want to receive on your phone, make sure the slider next to SMS is set to On.

3

Connecting with Friends

IN THIS PART . . .

Finding and adding friends

Sending messages and managing your inbox

Connecting and sharing with groups

Chapter **8**

Finding Facebook Friends

H undreds of sayings abound about friendship and friends, and most can be boiled down to one catch-all adage: friends, good; no friends, bad. This is true in life and also on Facebook. Without your friends on Facebook, you'll find yourself looking at a blank screen and asking, "Okay, now what?" With friends, you'll find yourself looking at photos of, say, a high school reunion and asking, "How did that last hour go by so quickly?"

Most of Facebook's functionality is built around the premise that you have a certain amount of information that you want your friends to see (and maybe some information that you don't want *all* your friends to see, but that's what privacy settings are for). So, if you don't have friends who are seeing your posts, what's the point in sharing them? Messages aren't that useful unless you send them to someone. Photos are made for viewing, but if the access is limited to friends, well, you need to find some.

On Facebook, the bulk of friendships are *reciprocal*, which means if you add someone as a friend, he or she has to confirm the friendship before it appears on both of your timelines. If someone adds you as a friend, you can confirm or delete the request. If you confirm the friend, *congrats!* You have a new friend! And if you don't, the other person won't be informed.

If you're low on friends at the moment, don't feel as though you're the last kid picked for the team in middle-school dodgeball. You can find your friends on Facebook in many ways. And if your friends haven't joined Facebook, invite them to join so they can be your friends on Facebook as well as in real life.

What Is a Facebook Friend?

You might be wondering, "What is a Facebook friend"?" Good question. In many ways, a *Facebook friend* is the same as a real-life friend (although, as the saying goes, "You're not real friends unless you're Facebook friends"). These are the people you hang out with, keep in touch with, care about, and want to publicly acknowledge as friends. Facebook friendships sometimes start on Facebook — through a group or post — but more often they are the people you know in real life already. Facebook friends are the same people you call on the phone; stop and catch up with if you cross paths at the grocery store; or invite over for parties, dinners, and other social gatherings.

In real life, there are many shades of friendship — think of the differences acquaintances, a friend from work, an activity buddy, and best friends. Facebook doesn't really have these nuanced differences. By default, all friendships are lumped into a blanket category of *friend.*

Here are the basics of what it means to be friends with someone on Facebook, though note that each of them comes with a few caveats on how it can be adjusted by either person in the friendship.

>> **They can see all the stuff on your timeline (such as your posts and other information) that you have set to be visible to friends.** Remember, this is what happens by default. You can control which friends can see which posts more specifically by learning about your privacy options (in Chapter 6).

>> **They see new posts you create in their News Feeds on their Home pages.** Again, the information your friends see in their News Feed depends on the audience you've chosen to share each post with. It may depend also on your friends' News Feed settings.

>> **You can see their posts and other information on their timelines.** Whether you see their posts depends on their privacy settings, but in general, you'll be able to see more as a friend than you did before you became friends.

>> **You see new posts from them in your News Feed on your Home page.** Whether you see their posts depends on your their sharing settings, but more importantly, you can control whose posts you see in your News Feed by

managing your own News Feed settings and preferences. See Chapter 4 for more information on News Feed settings.

>> **You'll be listed as friends on one another's timeline.** This is a small detail, but it's important in understanding the difference between becoming friends with someone and simply following someone. When you become friends with someone, anyone who navigates to that person's timeline will be able to see that you are that person's friend.

Adding Friends

Over time, Facebook has created some unique lingo. One of the most important Facebook terms is the verb "to friend." *Friending* is the act of adding someone as a friend. You may overhear people use this casually in conversation: "You won't believe who finally friended me!" And now you, too, will be friending people.

Sending friend requests

Now that you know what a friend is, it's time to send some requests, and maybe even accept some pending ones. Facebook enables you to send a friend request in two ways: the Add Friend button and the add friend icon, shown in the margin.

For example, if you search for *Carolyn Abram* in the search box in the top left, Facebook tries to autocomplete what you're typing, as shown in Figure 8-1.

Click the name of the person you think you want to add. The person's timeline appears, so you can verify (we hope) that this person is the one you want to add. Below the cover and profile photos are several icons. Click the blue add friend icon to send a friend request.

FIGURE 8-1:
The search results for Carolyn Abram.

If you decide that sending a request was a mistake, press the blue icon again to cancel the request.

Depending on people's settings, you might not see an add friend icon. If you can't send a friend request, you can usually send the person a message or follow her public posts without becoming friends first.

You won't be friends with someone until she confirms your friend request. After she confirms, you're notified by a red flag appearing above the notifications icon in the top bar.

So what does your potential friend see after you send a request? That is a brilliant segue into the next topic, accepting friend requests.

Accepting friend requests

When you receive a new request, a little red flag appears over the notifications icon (bell) in the top bar of each page. You may also be notified in your email or on your phone. Clicking this icon opens the Notifications dialog, as shown in Figure 8-2.

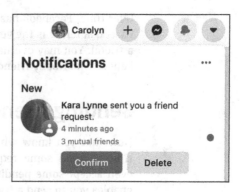

FIGURE 8-2:
Click Confirm or Delete.

To accept the friend request, click the Confirm button. You now have a friend. To reject the request, click Delete; you will have the option to mark that request as spam if you suspect it's from a fake profile.

If you want a little more information about the person before you accept her friend request, click her name to visit her timeline. Here you can explore a bit to jog your memory or make sure this person is someone you want to be friends with. Usually checking out a person's photos and friends list will be enough to figure out what you need to know. Oh, right, *that* Colin — the one with the hair who's friends with Penelope.

Some people worry about clicking that Delete button. If you're not sure what you want to do, you can always leave the request untouched. But never hesitate to click Delete for someone you really just don't want to be friends with. Facebook won't notify the person that you ignored the request.

Choosing your friends wisely

Generally, you send friend requests to and confirm friend requests from only people you know. If you don't know them, click Delete. Accepting a friend request from unknown people has a tendency to ruin the Facebook experience because it puts random content in your News Feed, exposes your own content to people you don't know, and is generally a bad practice. Remember the lecture you got about choosing good friends when you were in high school? It's every bit as true now.

If there are people you don't know personally but find interesting (such as a celebrity or public figure), you may be able to subscribe to their posts without becoming friends with them. (More on that topic in the "Following" section near the end of this chapter.) If that's not a possibility, you could add them as friends and then add them to your acquaintances or restricted friends list, if you use these lists to control your privacy.

Another common misconception about Facebook is that it's all about the race to get the most friends. This is very, very wrong. Between the News Feed and privacy implications of friendship, aim to keep your friends list to the people you care about. Now, the number of people you care about — including the people you care about the most and those you care about least — may be large or small. It doesn't matter how big or small your list is as long as the people you care about most are on it.

Finding Your Friends on Facebook

How do you get to the people you want to be your friends? Facebook is big, and if you're looking for your friend John, say, you may need to provide some more detail. Facebook has a couple of tools that show you people you may know and want as your friends, as well as a search-by-name functionality for finding specific people.

Checking out people you may know

After you have a friend or two, Facebook can start making pretty good guesses about other people who may be your friends. Facebook primarily does this by looking at people with whom you have friends or networks in common. In the People You May Know box, you see a list of people Facebook thinks you may know and, therefore, may want as friends. The People You May Know box appears all over the site — on the Friends page, on your Home page, and sometimes on your timeline. Usually the boxes include profile pictures, the potential friends' names, and some sort of info such as how many mutual friends you have or where the other person attended school. These little tidbits are meant to provide context about how you might know that person.

When you find yourself looking at the People You May Know list and you do, in fact, know someone on the list, you can add that person as a friend by simply clicking the blue Add Friend button. If you're not sure, you can click a name or picture to go to that person's timeline and gather more evidence about if and how you know that person. Then you can decide whether to add that person as a friend. If you're sure you don't know someone, or if you do know someone but are sure you don't want that person as your Facebook friend, click the X that appears in the upper-right corner. After you do that, she stops appearing in your People You May Know list. As you remove people from the list, more pop up to take their places. This fun can last for hours.

Browsing friends' friends

One easy way to build out your friends list when you're just getting started is to check out your friends' friends. Usually you have a friend or two who will already be friends with most of the people you want to be friends with as well. To browse your friends' friends, follow these steps:

1. **Click a friend's name in News Feed or search for a friend using the search box in the top bar.**

    The person's timeline appears.

2. **Click the Friends tab, located below their profile picture, name, and bio.**

    The Friends tab displays all your friends' friends, as shown in Figure 8-3. By default, you will see a list of that person's friends, with any friends you have in common listed first. You can use headings such as College or Current City to see this person's friends who share an alma mater or city with that person. Scroll down to see more friends.

3. **When you see a name you want to add, click the blue Add Friend button next to that name.**

    Your friend request is sent. When that person is ready, he or she can respond. We hope it will be the start of a beautiful Facebook friendship.

WARNING

Some friends may have used privacy settings to hide their friends list, in which case you won't be able to see their friends.

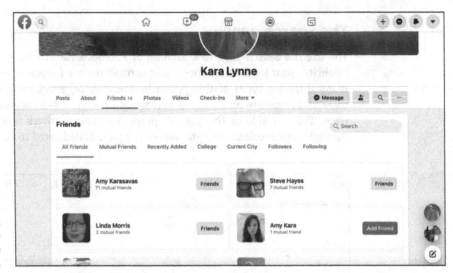

FIGURE 8-3:
A friend's friends
might just be
your future
friends.

Using the search box

The search box in the top bar lets you search a whole lot of things on Facebook: Pages, groups, events, and even things your friends have posted. But most of the time, you use it to search for people. You might search for people you're already friends with but you just want to go to their timelines. Sometimes you will search for people whom you aren't friends with yet but want to reach out to. In this section, we cover two types of search: basic searching using autocomplete, and using the Search page with filters.

Basic search

In the big bar on the top of any page on Facebook, you can click the magnifying glass icon to display a search box. The search box is where you type the name of a friend you may be searching for (or, for that matter, anything on Facebook you're be searching for). A list appears, displaying your most recent searches. If you want to go back to a previous search result, click that item in the list.

If you want to start a new search for someone, begin typing the person's name in the search box. Facebook will autocomplete as you type, so pay attention to the list below the search box as you type. Each letter you add changes the results. If you see a name with a profile picture and click that, you will be taken to that person's timeline, where you can add the person as a friend.

If you don't see your friend showing up as result, you need to use the Search page to complete your search. To go to the Search page, press Enter when you finish typing or click any results that has a magnifying glass icon. On the Search page, you can refine your search to find the person you're looking for.

The Search page

You use the Search page to search all of Facebook for anything — your favorite celebrity, your local coffee shop, your softball team's Facebook group. Because so many different types of things can be found here, Facebook offers filters that allow you to zero in on what you're looking for. Figure 8-4 shows a sample search page. The left side of the page displays the various filters you can use to sift through your results. The right side of the page is dedicated to those results.

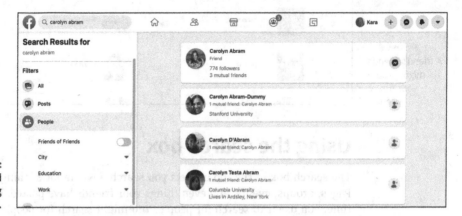

FIGURE 8-4:
Get to the friend
you want using
search filters.

When you land on the Search page, you're usually looking at all results. If you're looking for a particular person you'd like to become friends with, make sure to click People in the left sidebar. The right side of the page displays previews of people's timelines, usually with their name, profile picture, and any information about them that Facebook thinks will make them easier for you to identify. For example, it may show you where they are from, where they went to school, or if they have any friends in common with you.

If you still have a lot of results to sift through, you can narrow the results by using the filters within the People filter (on the left side of the page):

>> **Friends of Friends:** Move the slider from grey to blue to filter your search to only people who have friends in common with you.

>> **City:** Click the down arrow to use your hometown or current city as a filter in your search results. You can also enter any city name in the text box that opens.

>> **Education:** Click the down arrow to use any schools you attended as a filter on your search results. You can also enter the name of any school, even if you didn't go there.

>> **Work:** Click the down arrow to use any workplaces you've listed on your timeline as a filter on your search results. You can also enter the name of any company, even if you didn't work there.

REMEMBER

When you add more than one filter, you'll see *fewer* people because now Facebook is looking for people who both worked at Mom's Pizza *and* went to Hamilton High School. To see more results, use only one filter at a time.

When you've added a filter, you see that particular filter (for example, Wiley Publishing) displayed in blue on the left side of the page. To remove the filter, click the blue X next to it.

Managing How You Interact with Friends

After you do all the work of finding and adding friends, at some point, you may find that things are feeling a little out of control. Chances are you may be seeing posts from someone you find uninteresting; you might not be sure who, exactly, can see your own posts anymore; or you may just want to tidy up your friends list. At this point, it's a good idea to get acquainted with the way Facebook automatically helps you end the madness, as well as some of the specific actions you can take.

News Feed preferences

Often people with lots of friends will find that they have a few friends whose posts are simply not to their taste. It might be that they overshare, post too frequently, are too negative or too positive, or any number of other reasons. Whatever the reason, you can exert some influence on what Facebook shows you in News Feed preferences.

To get to News Feed preferences, click the down arrow (Account menu) in the top bar, and then select Settings & Privacy from the menu that opens. Select News Feed Preferences from the subsequent menu. The News Feed Preferences window appears, as shown in Figure 8-5.

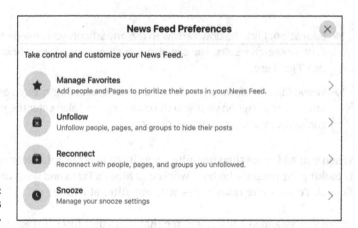

FIGURE 8-5:
Adjust your News
Feed Preferences.

The four options here enable you to keep News Feed more to your liking:

>> **Manage Favorites:** Designate people whose posts you always want to see. These people's posts will appear first in your News Feed when they are available.

>> **Unfollow:** Designate the friends whose posts you no longer want to see. When you unfollow people, you remain friends with them (so they can still see your posts), but you won't see their posts in your News Feed.

>> **Reconnect:** If you unfollow too many people, your News Feed might get a touch dull. This section allows you to re-follow anyone you realize you miss.

>> **Snooze:** Snooze is like a trial run for unfollowing someone. You can designate people who you want to snooze for a certain period of time. They'll reappear automatically after that time.

Chapter 4 includes more detailed information about the ways in which you can control your News Feed.

Following

Following is a way of saying that you really really want to see someone's posts. On some social media sites, following is the primary way of interacting with other people. On Facebook, following is baked into friending someone or liking (or following) a Page. However, you can follow someone you are not friends with. In fact, by default when you add a friend, you follow them, even before the person has accepted your friend request. The catch here is that following someone without becoming her friend means you only see her public posts in your News Feed.

Many people allow others to follow their public posts without requesting friendship. Following someone is as easy as — actually, it's easier than — adding someone as a friend. Navigate to that person and click the Follow button beneath their cover photo. To unfollow her, click that same button (it now says Following) to open a menu of options. Select the Unfollow option (the last item on the menu).

If you're someone who plans on posting lots of public updates or are a public figure (locally or nationally), you can allow people to follow you instead of becoming your friend (they'll also be able to add you as a friend, but you won't have to accept their requests for them to see your posts). Follow these steps to allow people to follow you:

1. **Click the down arrow icon (Account menu) in the upper-right corner of the top bar.**

A menu appears.

2. **Select the Settings & Privacy option, and then select Settings.**

The Settings page appears.

3. **In the left sidebar, select Public Posts.**

The Public Posts Settings page appears.

4. **Use the privacy menu in the Who Can Follow Me section to determine who can follow you.**

You can toggle between Public and Friends. If you choose Public, congrats! People can now follow your public posts. After you opt into this feature, more settings appear on the page. These settings allow you to specify how followers can find and interact with your timeline.

Unfriending

It happens to everyone: After a while, you start to feel like a few people are cluttering up Facebook for you. Maybe you feel like you have too many friends, or maybe you and a friend have drifted apart. Don't worry; Facebook friendships are not set in stone. You can *unfriend* people just like you friend people.

To unfriend someone, do the following:

1. **Go to the person's timeline.**

2. **Click the Friends button.**

   A menu appears. The last item in this list is Unfriend.

3. **Click the Unfriend link.**

   Take a moment of silence. Okay, that was long enough.

TIP

People aren't notified when you unfriend them, but people who care about you (that is, family and close friends) have a tendency to notice on their own that, hey, you're not in their list of friends anymore. This can sometimes lead to awkwardness, so it might be worth using your privacy settings to further limit these people's knowledge of your life *before* you unfriend them.

Lots of people go through periodic friend-cleaning. For example, after changing jobs or moving, you may notice that you want to keep in touch with some people from that chapter in your life; others, you just don't. Unfriend away.

Chapter **9**

Just between You and Me: Facebook Messenger

C hances are that you communicate with other people online. You may use email or an instant messaging program such as iMessage or Skype. If you have a smartphone, you probably check email and text messages on it as well. Facebook Messenger has similar functionality and integrates into all these programs. In other words, messaging on Facebook stitches together email, texting, group chats, instant messaging, and video calling with a Facebook twist.

One special component of Messenger as opposed to other systems is that you no longer have to remember email addresses, screen names, or handles. You just have to remember people's names. The other benefit is that your entire contact history with specific people is saved in one place. This is a mirror of real life, where we usually sort our conversations with people not into topics but by *who* we were talking to. You might text a friend to ask if he wants to meet up for lunch, then speak on the phone to coordinate a time and place, then go to lunch and talk the entire time, and then afterward send him that article you were trying to summarize. Instead of splitting all these interactions into discreet emails, texts, or phone calls, Facebook thinks of your communication with that friend as one long, on-going discussion that lasts the entirety of your friendship.

This chapter covers the basics of Messenger, including sending and receiving messages and navigating your Messenger inbox. It also touches on some of the more advanced parts of Messenger such as video chat rooms and integrations for kids.

Sending a Message

Figure 9-1 shows the basic New Message chat window. We opened this by clicking the new message icon, which can be found in the bottom-right corner of Facebook. The message window floats over Facebook in the background. Remember, Facebook makes no real distinction between chats (instant messages) and messages. Everything gets saved to your message history. Anytime you click a Message button or link from a friend's timeline or timeline preview, a new message window opens, with that friend's name in the To field.

The New Message chat window has only two fields for you to fill out: a To field and a message box where you type the text of your message. Unlike emails, this chat window has no spaces for CC, BCC, or a subject line.

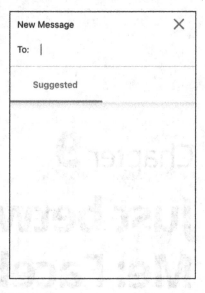

FIGURE 9-1:
The New Message chat window.

To address your message, simply start typing the name of the person you're messaging into the To field. Facebook autocompletes as you type, listing the name of friends, mutual friends, and then finally anyone with the name you're typing. When you see the name you want, highlight it and click or press Enter. You can type more than one name if you want to have a conversation with more than one person at the same time.

TIP

This isn't the only way to open a message window to a friend. You can click the Message button on a friend's timeline or preview to open a chat window. Clicking the friend's name in the Contacts section in the right sidebar also opens a chat window. Regardless of how you got here, messaging works the same way.

After you enter a friend's name, you can click in the text field to start typing your message. This field is at the bottom of the chat window, with *Aa* to indicate where you click to type. There are no rules about what goes here. Messages can be long or short, fat or skinny, silly or serious — whatever you have to say. Press Enter to send your message (or Shift+Enter to create a paragraph break). If an existing message thread was already started with that person, your new message simply gets added to the bottom of the conversation.

TIP

When you click the text box to type, you may notice that the icons at the top and bottom of the message window change from grey to blue. If you have a lot of chat windows open, this color change is an easy way to double-check that you are typing in the correct one.

As you and your friend message back and forth, you can continue to use Facebook — scroll up and down in your News Feed, browse photos, check out a timeline. Regardless of where you go on Facebook's site, your chat windows remain open at the bottom of the screen. They won't close unless you close them or close Facebook.

TIP

You can react to specific messages that your friend has sent by hovering your mouse cursor over the text and clicking the grey smiley icon that appears next to it. This opens a list of the usual Facebook reactions: love, laugh, surprise, cry, angry, thumbs up, and thumbs down.

Figure 9-2 shows a chat window between friends. It's designed to look somewhat like the text message interface on your phone. A profile picture of your friend is shown on

FIGURE 9-2:
A conversation between friends.

the left side of the chat window, next to the speech bubble that contains her message. Your own messages appear in color anchored to the right side of the chat window. You can scroll up to see older messages.

Sending a group message

You can message more than one person at a time. Doing so creates a new conversation among all the people you message. Everyone can see and reply to the message. So if you send a message to Daphne, Eloise, and Penelope, a new conversation is created. That cluster of three is now a group. When you're looking at that conversation, you can see all the messages that have been sent by all the people involved. As you're reading, you can see who said what by looking at the name and profile picture identifying each message. Each message is separated and has a timestamp so that you can see when it was sent.

The main thing to remember about group conversations is that you can't reply individually to members of the conversation. When you reply, all members of the conversation see your reply. If you're in multiple conversations with some of the same people, double-check to make sure you're in the right conversation before pressing Enter!

Group messages have some specific options, such as removing yourself from a conversation and naming a conversation. We cover those in the "Managing Messages" section.

Sending a link

If you want to add a link to a website or article to a message, you can copy and paste it into the message box. Facebook then generates a preview of the article so that your friend has more info before clicking the link. You can remove this preview by clicking the X in the upper-right corner of the preview. If you share a link from another website (many news sites have links to share articles, for example), you can select to share it in a private message and accomplish the same thing: sending the link and a preview to a friend, as well as explaining why you're sending it.

TIP

Much like when you create a post, you can make your messages a little less messy by sending only the preview without sending a long, unwieldy URL. After you can see the preview, simply delete the URL from your message. Your friend will be able to click through on the preview alone.

Sending a photo

To add photos to your message, click the picture icon (framed picture of a mountain) at the bottom of the chat window. Doing so opens an interface for navigating your computer's hard drive, so make sure you know where your photo is saved. If you want to share a photo that's already on Facebook, navigate to the photo, click the share link at the bottom of the photo, and share it in a private message.

REMEMBER

If you're sharing a photo on Facebook, privacy rules may sometimes prevent your friend from being able to see it.

Sending a sticker

Stickers on Facebook are a digital version of the stickers you might have used to adorn a school notebook or a letter from summer camp. An almost infinite number of sticker options are available for you to send, from smiley faces to ones related to various holidays to ones created by specific artists.

Clicking the square smiley icon at the bottom of a chat window opens the Search Stickers window, shown in Figure 9-3.

You can browse through the various categories of stickers here. Click a category name to see stickers in that category, and then scroll down to see the full collection. As soon as you click a sticker, it is sent to your friend.

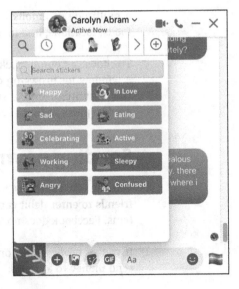

Along the top of the Search Stickers window are recently used sticker collections you have used. You can also click the plus sign here to go to the sticker store, where you can browse through sticker options and add them to your personal collections. After you have added a sticker collection (by clicking the Free button), close the sticker store by clicking the X in the upper-right corner. Then, when you reopen the Search Stickers window, you'll see your new sticker pack along the top.

FIGURE 9-3:
Selecting stickers.

Sending a GIF

A *GIF* is a file format that supports animated images. When people talk about GIFs, they're usually referring to how GIFs are used on the Internet: as ways to share a clip, usually on a loop, that captures a moment, an emotion, or a thought. So, for example, if someone tells you something that exasperates you, you could send a GIF of a famous person rolling his eyes to communicate that you're rolling your eyes as well.

You can click the GIF icon at the bottom of the chat window to browse through GIFs that you can send to a friend. By default, Facebook shows you trending GIFs (GIFs that other people have been sending). You can search for a type of GIF by using the search box at the top of the GIF list. As soon as you click a GIF, it is sent to your friend.

Sending an emoji

Emojis are small digital images that express an emotion or idea. The most common emojis are the yellow smiley (or frowny or crying) faces. You can select from a multitude of emojis on Facebook, ranging from the usual smileys to specific foods, locations, and modes of transportation.

To choose an emoji, click the round smiley icon at the bottom of the chat window. A window for browsing emoji appears. Click the one you want to add to any text you've written. Unlike stickers and GIFs, emojis are added in your text, so you need to press Enter to send them.

Sending payment

You can use Facebook to send money to friends. It requires both you and your friends to enter debit card info into Facebook, which it stores securely in its systems. Facebook does not charge a fee to send payments between friends.

If you've never sent money before, follow these steps the first time you want to send money to a friend:

1. **Click the blue plus sign in the bottom of the chat window to display more options.**

   These options are the same as the ones already covered, with two additions: sending an attachment (paperclip) and sending money (dollar sign).

2. **Click the $ icon in the bottom of a chat window with your friend.**

   A payment interface appears, shown in Figure 9-4.

3. **In the payment field, type the amount of money you want to send.**

4. **(Optional) In the Add a Note field, type the reason for the payment.**

5. **Click the Pay button.**

   A window opens for entering your debit card info for making the payment.

6. **Enter your debit card or PayPal info as prompted.**

   You can't use credit cards.

7. **Click the Pay button to confirm the payment.**

   The money is sent to your friend, who will need to enter debit card or PayPal info into Facebook to receive the money.

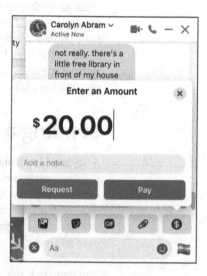

FIGURE 9-4:
Sending money to friends.

Once you send money, you can't cancel the payment. Be sure that you are okay with the money leaving your account before you click the Pay button.

If you'd like to add extra security around payments, check out the Payments section of the Settings page.

Sending an attachment

Much as it does in many email programs, an icon shaped like a paper clip signifies attaching files to a message. Clicking the attachment icon opens an interface for searching and selecting files from your computer's hard drive. You can attach photos, videos, documents, and so on.

Sending an instant emoji

In the example chat window shown in Figure 9-2, a rainbow flag emoji appears to the right of the text field for entering text. Click the flag and it is sent instantly to your friend. This instant emoji changes depending on the conversation you are a part of and the last emojis you used in that conversation. Sometimes the instant emoji may appear as the Facebook like icon and other times as other emojis. It's just a quick shortcut. You can edit what emoji appears as your instant emoji from the options menu described in the next section.

Starting a video or voice call

Assuming your computer is up to the task, you can click the video camera icon or the phone icon to start a video or voice call, respectively, with your friend (or multiple friends).

The first time you use either of these features, you may have to set up video calling, which may include installing Flash or another driver to your computer's hard drive.

When you initiate a video call, a pop-up window opens on your screen. Your friend will see something similar on her screen, asking if she wants to accept a video or voice call. If she wants to take the call, she simply clicks Accept and the two of you will be able to speak face to face (or voice to voice).

Video and voice calls can also be made from smartphones by using the Messenger app. You learn more about Facebook's suite of apps for phones in Chapter 7.

REMEMBER

Video calls assume that both people have webcams built-in or installed in their computers. If you don't have a webcam, you won't be able to make or receive video calls.

Managing Messages

When you send a message, chance are you'll get a reply pretty soon. When you receive a message, you see one of two things, depending on your settings and whether you were logged in to Facebook when the message arrived. You'll see either a new chat window (as if you're receiving an IM) or a little red flag over the Messages icon in the top bar — click the flag to open the inbox preview, and then click the message preview to open the conversation in a chat window at the bottom of your screen.

At the top of each chat window is a bar displaying the name of the person you're chatting with and several icons. To close the chat window, click the X. Closing a chat window won't lose any of the information in the chat — all your messages will be waiting for you the next time your friend messages you (or you message her).

To minimize a chat, click the horizontal bar icon. This keeps your chat close at hand without blocking your view of News Feed or any other parts of Facebook you may be looking at. When a chat window is minimized, a small thumbnail photo of your friend appears at the lower-right corner of your screen. If at any point you want to get back to messaging your friend, click that photo and your message window reopens. If a new message comes in from your friend, and small red flag will be added to your friend's thumbnail photo.

In the chat window, click the tiny down arrow next to your friend's name to view a full menu of options related to your conversation. Some of these options appear only in conversations with one friend. (We added a "one-to-one" note to those) and some appear only for group conversations. Most, however, appear for one-on-one and group threads alike:

>> **Open in Messenger:** Display your Messenger inbox and get more space on the screen to view your conversation history as well as search in it. More details about Messenger and its inbox can be found in the next section.

>> **View Profile:** See the profile of the person you're chatting with.

>> **Color:** Open a window to change the color of the word bubbles in the conversation. Click a color to select it. Facebook changes the color for all participants in the thread. This option is helpful when multiple conversations are happening at once.

>> **Emoji:** Select the emoji you want as your instant emoji for this conversation. Each conversation can have its own instant emoji.

>> **Nicknames:** Make sure everyone in the conversation knows who is speaking, especially if people use a variant of their name as their Facebook name. Nicknames are most helpful in group message threads. When you click this option, a window appears with a list of every participant's name. Click the name for which you want to add a nickname, type the nickname, and click Save. Keep in mind that all members of the conversation will see the nickname, so be nice!

>> **Create Group (one-to-one):** If you want to add a person to a conversation with a friend, choose this option. A new message window appears with the current friend's name prefilled. You can add more names and start a group conversation. Keep in mind that the message history from you and your original friend won't be copied over to the new thread.

>> **Conversation Name (group):** Give conversations amongst friends a helpful nickname such as Vacation Planning or Dummies. Select this option and a small text field appears at the top of the chat window. Enter the new name of your conversation there and press Enter.

>> **Members (group):** Open a box where you can see everyone who is part of the conversation, which is useful if you're part of a large group. Click the three dots icon next to any name to block that person or message them individually. If you are the original creator of the group, you also have the ability as an *admin,* or administrator, to eject someone from the group or make other people admins as well.

>> **Add Members (group):** If you're part of a group chat and you realize someone's been left out, you can add them by choosing this option.

>> **Leave Group (group):** If you don't want to be a part of a particular group conversation, you can simply to leave the conversation. Leaving a conversation posts a small notice to the rest of the members that you have left.

>> **Mute Conversation:** Muting a conversation allows you to stop receiving notifications (such as a flashing chat bar or a red bubble in the top bar) for a particular conversation. You can choose how long you want a conversation to be muted — as short as 30 minutes or as long as forever. We've found muting particularly helpful with group conversations, where we want to be able to view the thread when it's convenient but don't want to be distracted by a lot of people replying all at once.

>> **Ignore Messages (one-to-one):** Ignoring messages is a variant on muting messages from a particular friend. If you choose this option, Facebook will begin to treat the messages from that friend as if they were messages from a stranger. They'll show up in the Message Requests section of your inbox instead of in the main section. Your friend won't be told that you've ignored his messages.

>> **Block (one-to-one) or Block a Member (group):** You can prevent someone from sending you messages by choosing to block messages from her. If people are harassing you on Facebook, you might also consider blocking them or reporting them for harassment.

>> **Delete Conversation:** Deleting a conversation is permanent. It deletes the entire history of your messages with a friend. Keep in mind that deleting the conversation deletes it only for you. Your friend (or friends) will still be able to see the message history.

>> **Something's Wrong:** If you're getting odd messages from a friend promoting something he wouldn't normally promote, his account might have been *phished*, meaning someone has gained access to it. Report the spam messages to protect yourself, your friend, and other users from having the same thing happen to them. You can also report messages for being harassing or hateful. If you're reporting those actions, you may also want to consider blocking the person sending the hateful content to you.

USING FACEBOOK PORTAL TO CONNECT

Portal is Facebook's stand-alone video-messaging device. Unlike in the rest of Facebook, you have to pay for it. Once you have it, you log in with your Facebook information and can use it to video chat with your Facebook friends. You don't *need* Portal to take advantage of Facebook's video-chatting features, by any means, but if you find yourself video chatting with the same people regularly, you might find that Portal enhances the experience.

The camera on Portal uses an algorithm to track movement and keep the camera focused on the person speaking. It can zoom in and out to accommodate more people. It can track movement, so if a child runs across the room, the people on the other side can watch the child for longer than the quick blip a regular camera would allow. It comes with various games built in, so if both people chatting have Portal, they can compete in silly games where you, for example, move your head in the video to try and catch a digital donut in your mouth. Over time, Portal has added more features such as the capability to use it with Zoom and other video-chatting services.

Carolyn's family distributed Portals to various long-distance family members during the COVID-19 era, and it provided a great way to connect with grandparents, aunts, uncles, and cousins. It sat on the kitchen counter, where the kids might eat lunch while chatting with their grandparents, or show off the silly effects that turned their faces into dragon heads. When not in use, it was a digital picture frame for a variety of her Facebook photos.

Checking Out the Chat List

The chat list, shown in Figure 9-5, is like a buddy list for instant messaging. It's on the right side of the page.

By default, the Contacts section of the chat list displays the friends you messaged with most recently. A green dot next to a name means the person is currently on Facebook and will likely see a message if you send one. A green timestamp such as 20h or 2d indicates how long it's been since they've been on Facebook. Rest assured that any messages you send to them will be delivered. Often people receive notifications in email or on their phone when they get a new Facebook message, so just because they haven't been active recently doesn't mean they won't get your message soon.

FIGURE 9-5:
The chat list.

Below the friends you message most often are the rest of your friends, as well as any existing group conversations you might want to participate in. Scroll down to browse your friends.

To quickly find the friend with whom you want to chat, or to see if that friend is even online, click the search icon next to the Contacts header at the top of your list of friends. Start typing in the search box that opens. As you type, the list of online friends narrows to those with names that match what you've typed. After you see the friend you were looking for, click the friend's name to open a new message window to that person.

Click the three dots icon next to the search icon to open the Chat Settings menu. This menu offers the following options for adjusting your Chat experience:

>> **Incoming Call Sounds:** Decide if you want your browser to make a ringing noise when someone calls you using Facebook.

>> **Message Sounds:** If this option is selected, you will hear a sound every time you receive a new chat.

>> **Pop-up New Messages:** If you don't like messages suddenly opening at the bottom of your screen, you can turn off this setting. You'll see a red flag appear over your messages icon in the top bar when you receive a new message.

>> **Show Contacts:** If you don't like seeing the list of friends in the chat list, you can hide them by turning off this setting. The right column will still be there, you just won't see your friends listed in it anymore.

>> **Turn Off Active Status:** This option hides the fact that you're currently active on Facebook (or Messenger) from your friends. If you're trying to do something on Facebook without people bothering you simply because you are active and available, this option will help cut down on the number of incoming messages.

>> **Block Settings:** Selecting this option brings you to the Blocking section of the Settings page. There you can block people from messaging you or from interacting with you at all on Facebook.

Navigating Messenger

After you're comfortable sending and receiving messages to and from your friends, it's time to find out about your Messenger inbox, where all your messages are collected for easy viewing at any time. Messenger is also the name of the mobile app you can download and use on your phone to communicate with Facebook friends. You learn more about the Messenger app in Chapter 7.

Messenger is organized a bit differently from traditional email inboxes. Most significantly, messages you receive from people you aren't friends with and are unlikely to know are separated from the conversations you're having with friends.

To understand how this works, look at how Messenger is organized on the page. From your Home page, click the Messenger icon (word bubble with lightning bolt) in the top bar. You see a preview of your most recent messages. At the bottom of the preview, click See All in Messenger to go to Messenger. Figure 9-6 offers a snapshot of a sample Messenger inbox.

The left side of the page displays your conversations. Each conversation gets its own line in the inbox. Like your email inbox, these conversations are organized from most recent near the top to older ones toward the bottom of the page. As you scroll down, Facebook will continue to load your conversation history.

The main portion of this page, the center area, is where conversations appear. As you click different conversations on the left, the contents of that conversation — messages, photos, links, and files — appear in the main portion of the screen.

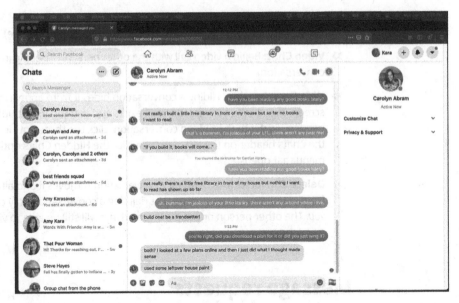

FIGURE 9-6:
Welcome to your
inbox.

Figure 9-7 shows a close-up of two conversation previews. The top one is a group message, the bottom one is a one-on-one conversation with a single friend. Your friends' names are listed, and their profile pictures are displayed. On the right side of each is the conversation's timestamp. And below the name of the person or people you're talking to is a snippet of what was most recently said. Messages that you haven't read yet will appear in bold.

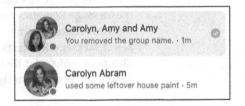

FIGURE 9-7:
Conversation previews in the Messages inbox.

When you hover your mouse cursor over a conversation, a three dots icon appears to the right of the names and other text. Clicking this icon opens a menu with options for that conversation:

>> **Mark as Unread:** Much like email, mark a conversation as unread if you'd like to return to it later.

>> **Mute Conversation:** Muting a conversation means you won't receive notifications about incoming messages for the selected amount of time.

>> **View Profile (one-to-one):** Visit the profile of the person with whom you are messaging.

- **» Audio Call:** Begin an audio call with the other person (or people).

- **» Video Chat:** Begin a video call with the other person or people in that conversation.

- **» Hide Conversation:** Hiding a conversation hides it from appearing on this screen, though the conversation still exists and you can look it up later if you need to. To get to the hidden conversation, click the three dots icon next to the Chats header on the left, and then click the Hidden Chats option from the menu that opens.

- **» Delete Conversation:** Deleting a conversation removes it and all its contents from your inbox. Keep in mind that deleting a conversation only deletes it for *you*. The other person or people involved in it will still be able to view it in their own inboxes.

Message requests

Whenever you get a message from someone you aren't friends with yet, it goes to the Message Requests section of your inbox. To get to your message requests, click the three dots icon next to the Chats header in the left column of Messenger. Select Message Requests from the menu that opens. The Requests page appears. This page looks a lot like your Messenger inbox, except all the messages are from people you aren't directly connected to.

USING MESSENGER ON THE DESKTOP

If you like using Messenger and all its features but you don't like being distracted by your News Feed, you can download a desktop app for both Mac and Windows computers. When you use the Messenger Desktop app, it opens a stand-alone program for using Messenger.

To download Messenger for your desktop, go to the Windows App Store or the Apple App Store, depending on which device you use. Search for Facebook Messenger and then click Get. Download and install the app as directed by the site. Then open it and log in with your Facebook credentials.

The Messenger app on a Mac resembles most other messaging apps: You can see a list of conversations on the left, and then focus on a selected conversation on the right. The app should look very familiar if you've spent time in the inbox on Facebook, or if you've used Messenger on your phone. You can use video chat, make voice calls, and create and join rooms, all without needing to log in to Facebook on your browser.

New messages appear in bold. Click any message to read its contents and decide whether you want to accept the message. If you want to accept the message, simply type a reply; the message will be moved to your regular inbox. If you don't want to accept it, click the I Don't Want to Hear from *Part* link just above the reply box.

If you decline a connection request, the person is not notified, but the person's message is permanently deleted from Messenger. You can also leave connection requests in a sort of limbo, where you neither respond, accept, nor decline the request. It will just hang around in the Message Requests section of your inbox gathering dust until you are ready to do something with it.

TIP

Because message requests go to a separate inbox, you may find that messages you send to people you're not friends with might not be seen or replied to. If you're messaging someone from a post in a group you both belong to, for example, leave a comment to let the person know you sent a message. People usually say something like "I just sent you a PM," where PM means *private message*.

Conversations in the inbox

The center portion of the Messenger inbox is dedicated to whichever conversation you have selected. The most important thing to notice is that all the content here is the same content you would see in a chat window. Facebook doesn't care where you wrote messages — they all go into your message history. The most recent message is on the bottom of the page. Scroll up to see older messages.

At the bottom of your conversation, below the most recent message, is the message composer. The message composer is similar to the one you use in the chat window. Simply type your response and press Enter to send. You can click the icons for sending photos, stickers, GIFs, emojis, money, and instant emojis.

To the right of your conversation is an info column about the conversation you're looking at. You see info about the person (or people) you're talking to, customization options, privacy and support options, and photos you've shared in the conversation.

Most of the options in the Customize and Support sections are the same as those in the chat list. The only notable new setting is Search in Conversation, in the Customize Chat section. Because Facebook messages aren't broken into discreet emails with unique subjects, trying to scroll through a long message thread to find information such as where you guys were supposed to meet or the identity of a mysterious newsletter publisher can be challenging. You can use search to try to find this type of information.

Clicking Search in Conversation opens a search box at the top of the message thread. Simply type the term you're searching for and press Enter. Facebook will display the most recent use of that term, highlighted with the messages that were sent immediately before and after that term was used. Use the arrows next to the search box to flip to the next occurrence of the term. When you've finished using Search, click the Close button to the right of the search box to return to the entire conversation.

Messenger settings

To access your Messenger settings, click the three dots icon next to the Chats header at the top of the left column. This opens a menu with the following options:

>> **Preference:** Open a pop-up window to access a few settings and links that may be helpful to you as you use Messenger. You can toggle your online status on and off, and enable or disable sounds. You also have links to manage your payments and blocking settings. You can also choose your emoji's skin color.

>> **Active Contacts:** View a list of all your friends who are currently active on Facebook.

>> **Message Requests:** View all connection requests that you have not responded to yet.

>> **Hidden Chats:** View conversations you've archived.

>> **Help:** Go to the Facebook Help Center.

Getting into Rooms

Facebook rolled out its Rooms feature right around the time COVID-19 adjustments were being made all around the world. People were beginning to spend more time online together and happy hours, birthday parties, and other social events began happening in virtual spaces instead of real ones. Rooms is Facebook's version of many other digital hangout spaces such as Zoom, Microsoft Teams, and Google Hangouts.

WARNING

To use the Rooms feature, you must be using the Messenger app (for either your phone or desktop) or the Google Chrome web browser. If you're using Safari or Firefox or another common browser, you will not be able to enter rooms created by you or your friends.

To create a room of your own, follow these steps:

1. **Click the + icon in the top bar, and then choose Create ⇨ Room.**

   The Create Your Room dialog appears.

2. **Choose a name for your room.**

   By default, your room is named after you. You can select from other options or create your own by clicking Room Name. This allows you to browse other options such as Here All Day and Happy Hour. Click +New to use a custom name, such as V. Woolf Book Club.

3. **Select a start time.**

   By default, a room opens immediately after you click Create Room. If you'd like it to start at another time, click Start Time, and select a date and time when you'd like your room to become available.

4. **Decide who can see that your room is open.**

   By default, rooms are set to be visible to only friends you invite. You can make it so that all your friends can see that you've created a room, so that they can pop in and out as they see fit.

5. **Click the blue Create Room button.**

   The Create Room box shifts to show you the details of your room, including a link you can send for people to join the room.

6. **Send invites to friends whom you'd like to join your room.**

   Tap the Send button next to any friend's name to invite that person to join the room.

7. **Click the blue Join Room button.**

   The room opens in a new browser tab or window. You can check out your video and audio settings before clicking the Join as *your name* button. After you've joined, enjoy your time in virtual space with your friends.

Facebook rooms have many of the same functions as most video-chatting services: You can mute your audio or your video, see the participants, share your screen, and hang up when you're done. Additionally, Facebook has some special options, found in the upper-right corner of the room:

>> **Play Games:** This shortcut allows you to find games built by outside developers that you can play against the people in your room. You can play a variety of games, including party game classics such as Charades and Scattergories.

>> **Watch Videos Together:** You can browse videos that will be played at the same time for everyone in the room. This is a convenient feature for watching things together without having to worry about complicated audio echos — Facebook makes sure everyone can hear the same thing at the same time without blowing out your eardrums.

>> **Live Video:** If you want to broadcast more widely the discussion happening in your room, you have the option to turn it into a live broadcast.

>> **Open Chat:** If you want to chat via text as well as video, you can start a group chat that everyone in the room can view and add to.

TIP

The room still exists even after you leave. It can't be used by your friends unless you're there, but if you want to meet up regularly, simply rejoin the room you created in the past. All details (name, invitees, and so on) remain the same. Find rooms you've created in the past at the top of your News Feed, below the share box.

Messaging on the Go Using the Messenger App

Facebook integrates its messaging system seamlessly with its smartphone apps. It even has an app just for messaging: Facebook Messenger. Regardless of where you're looking at a message — on your phone, in a chat window, or in the inbox — you see roughly the same thing, with only slight adjustments to account for the differently sized screens.

The Facebook Messenger app is pretty simple. It displays the contents of your inbox: a list of conversations, with the most recent at the top. Tap any conversation to open and read it. To reply to a conversation you're viewing, tap the text box at the bottom of the screen to open a keyboard for typing a new message.

To start a conversation, go to your inbox and tap the new message icon (pencil inside a box) in the upper-right corner of the screen. The New Message screen appears, and you enter the names of the people you want to message and the message itself. This should feel eerily similar both to sending a text message and to sending a message on Facebook.

If you're using an app such as Facebook Messenger on your phone, you will likely be notified on your phone each time you get a new message. You can adjust these settings from the app itself. If you want to know more about using Facebook from your mobile phone, check out Chapter 7.

Messenger Kids

Most of Facebook is not for kids. You need to be over 13 to use Facebook; that is simply the rule. However, if you're a parent of children under the age of 13, you can create a special type of account for them that allows them to use a version of Messenger to communicate with (approved) friends on their own. Messenger Kids is a stand-alone app that you can download to a phone or tablet that your child is allowed to use. After it's downloaded, you need to log in with your Facebook credentials. After that step, you can create profiles for your children. These profiles are not Facebook accounts and they do not have the same capabilities as your Facebook profile, though you will add your child's name and birthday.

After you create the accounts for your children, you get to choose who else they can chat with using Messenger Kids. Facebook recommends any of your friends' kids who have profiles, and gives you a chance to connect any of your friends that you'd like to be able to chat directly with your kids.

With Messenger Kids, parents make the decisions about how their kids can connect with other kids and grown-ups. If you have a mature older kid, you can give them the ability to add their own contacts. Or if you prefer, you can make it so that you must approve all new connections. One useful feature is that Facebook creates a unique code for your child that you can give to another grown-up when you're trying to connect your kid to school friends whose parents you are not Facebook friends with.

When your children have a few chat connections, they can message back and forth with their friends, play games with them, share photos that they take from the app, and generally connect. Facebook doesn't allow kids to delete messages that they've sent or received, so as a parent you will always have the ability to monitor what's been said and whether anyone new has been trying to connect with your child.

IN THIS CHAPTER

» Joining and leaving a group

» Sharing and talking with group members

» Adjusting group notifications

» Creating groups

» Managing groups and group members

Chapter **10**

Sharing with Facebook Groups

B y now, you've probably gained some Facebook friends. We hope you've also started sharing with them by posting statuses, photos and links, and you're seeing your News Feed fill up with much of the same. All that is great, but as you share certain things, you may find yourself thinking, "Only people from work will care about this article" or "Really, only the people in my book club will find this funny." What you need is a designated space to hold these conversations. A *group* creates this space by fostering interactions with a specific group of people, almost as though you were sitting in the same room.

Groups can be large or small. They may have very active participants or people who sit back and observe. Some groups may involve ongoing forums; others may exist only to achieve a goal (for example, planning a big event). Groups can be open to anyone in the world to join, or they can be private affairs that require invites to join. Whatever forms groups take, they are a great way to achieve a sense of community. This chapter covers how to use the features of groups, as well as the many options that come with using, creating, and managing groups.

Evaluating a Group

When someone adds you to a group, a notification is sent to your Facebook Home page. The next time you log in, you'll see a red circle over the notifications icon (the bell icon in the top bar). You may also see a post about the invite in your News Feed. The notification will tell you the name of the group and who added you; click the notification to check out the group and make sure you want to be part of it.

REMEMBER

When you receive an invite, you're able to see the group in preview mode. *Preview mode* allows you to try on the group for size before you make the decision to join or leave the group. You may begin to see group content in your News Feed and interact in a limited capacity with other members by, say, reacting to a post. You can't comment on posts or invite members yourself. Other members can see that you're in preview mode, and you do not yet count as an official member of the group.

Preview mode expires after 28 days, after which you will no longer see group content on your News Feed. If you interact with any of the group's content before preview mode is up, it will extend another 28 days. If you want to exit preview mode sooner, click Decline Invite (at the top of the group's Home page) to remove yourself from the group or click Join Group to accept the invite and become a fully fledged member.

Before accepting an invite or clicking the Join Group button, you'll want to explore what the group is all about. The most important thing to consider is whether or not you want to be part of that group. It might be tempting to accept a group invitation, but you probably don't want to join groups that advocate for causes you don't believe in or are about something you don't find interesting.

Luckily, information is available to help you on your quest, as you can see by the sample group shown in Figure 10-1. Groups are designed to look a bit like a timeline — the cover photo, share box, and recent posts from group members in the center of the page should all look familiar to you. Other parts of the group are unique, such as the About and Recent Media sections.

A group's privacy and visibility settings tell you a lot about how that group operates. In Figure 10-1, the privacy and visibility of the group are listed in the About section on the right of the page. Facebook Groups have three privacy and visibility options:

>> **Public and Visible:** A public visible group is visible to everyone and available for anyone to join. In other words, anyone who uses Facebook can see the posts and the members of that group.

FIGURE 10-1:
Checking
out a group.

>> **Private and Visible:** A private visible group can be found by anyone on Facebook, but only members can post in the group and see other members. You will be able to see the admins of the group as well as the About section and any group rules. People are added to private groups by other members, or they can request to join.

>> **Private and Hidden:** A private hidden group is essentially a secret. Only members can see that the group exists, who its other members are, and the posts that have been made. If you're invited to join a private hidden group and you choose to leave it, you won't be able to join again later, because you won't be able to find it.

If you're unsure about joining a group, check out its purpose, the activity level, and any rules for being a member. For example, a group about snowboarding in New York might say something like "A place to share tips, arrange carpools, and review East Coast mountains." If what you want is info about snowboarding in California, you know to be on your way.

If the group is a public group, also check out the top posts in the center of the page by clicking the Discussion tab under the group's cover photo. These are posts that group members have shared directly with the group and not on their personal timelines. In Amy's houseplant group, for example, many people share photos of wilting plants asking for advice on how to get droopy leaves looking perky again. Use one of the options for sorting posts if you'd like a more comprehensive overview. Click the grey arrow next to New Activity and choose from Recent Posts (the newest posts) or Top Posts (posts with the most comments and reactions). Looking at a group's posts tells you a lot about what to expect from the group. Are posts relevant to the group or not? Is there a lot of discussion or not? Are you interested in the posts you're seeing? These are the sorts of questions to mull over when deciding whether to join a group.

Finally, ponder the size of the group itself. Facebook groups might be enormous or very tiny. You might be interacting with a handful of people you know or thousands of people you have never met and never will meet. Facebook has slightly different features for big and small groups — whether it's the tools for managing large groups or the fact that in smaller groups, you can see how many people have seen a particular post. Make sure you're comfortable sharing with the number of people in the group that you want to join.

After you acquaint yourself with the group, you can decide whether you want to remain a member or leave the group.

If you decide the group is for you, click the blue Join Group button. Depending on the group's settings, you may be prompted to answer up to three questions to help admins determine your eligibility. You may also need to select the box that says I Agree to the Group Rules. Group rules tend to cover what type of behavior the admins encourage or discourage. For example, Amy's houseplant group asks members to refrain from self-promotion. Click the blue Submit button to finish your request to join.

REMEMBER

Even if you aren't prompted to fill out a questionnaire, keep in mind that the group admins may need to approve your request to join before you officially become a member. This process may take a few days, so don't despair if you're aren't immediately admitted to the group of your dreams.

If you decide the group isn't for you, follow these steps:

1. **Do one of the following:**

   - *If you were invited by a friend:* Click Decline Invite, and then click Decline in the pop-up window that appears. You're no longer part of the group.

   - *If you weren't invited by a friend:* Click the three dots icon at the bottom of the cover photo. In the drop-down menu that appears, choose Leave Group and follow Steps 2 and 3 to complete the process.

2. **Decide whether you want to prevent members of the group from sending you another invite.**

   If you don't want another invitation to the group, click the Prevent People from Inviting You to Join This Group Again option. You'll never be added to the group again. If you think you might want to be part of the group in the future, leave this option deselected.

3. **Click Leave Group.**

If you decide instead to stay in the group, you're ready for the next section, which describes sharing with fellow group members.

TIP

If you want to leave a group because you're being inundated by notifications about new posts, you can instead just turn off notifications for that group. Check out the "Controlling notifications" section of this chapter for details.

Sharing with a Group

The whole point of creating or joining a group is to enable communication, so let's get started! Ways to get involved include posting to the group, commenting on others' posts, chatting with group members, and creating files or events.

Using the share box

You post to a group the same way you post to a timeline or the News Feed: by using the share box. When you navigate to the Discussion tab, the share box appears above the recent posts. The share box here works the same as the share box in News Feed, with a few extra options that we go over in this chapter.

REMEMBER

When you share something to a group, you're sharing it with only the members of that group. Keep in mind that you may not be friends with everyone in the group and that you may be sharing with many people who typically couldn't see the things you post.

Although the share box works almost the same way across Facebook, we briefly explain some options here in the context of groups.

Writing a post

Posts are basically status updates that you share with the members of a group (unless the group is public, in which case anyone can see your post). You might post an update just to say "Hi" or to start a discussion with group members. To write a post, follow these steps:

1. **Click in the share box (What's on Your Mind, at the top of the group page).**

2. **Type whatever you want to say.**

   For a houseplant group, you might write "How often should I water my cacti?" or "Does anyone know where I can buy seeds online?" You can also post a link to a relevant article or website in this space. You can see a post in progress in Figure 10-2.

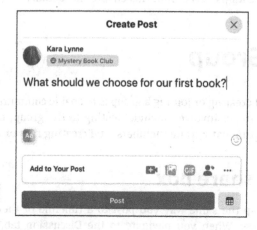

FIGURE 10-2:
What's on
your mind?

3. **(Optional) Add photos, videos, GIFs, tags, activity, or location information to the post.**

   Click the three dots icon at the bottom of the share box to see the many ways you can customize a post.

4. **Click Post.**

   Your post appears in the group, and group members can see it in their notifications and News Feeds.

TIP

If you want to share a link, usually an article, a video, or other online content that you want the group to see, simply type or paste the complete link, along with your thoughts or opinions, in the share box. To keep the post looking neat, delete the link from the text box after you've typed it. The shared content remains on the post without the messy URL address.

Creating a poll

The Create Poll feature is handy for groups. Polls allow group members to gather information in a more efficient way than simply posting a question and sorting through the comments for answers. Members can vote on polls from the Recent Posts section of the group. To ask a question and poll your group, follow these steps:

1. **Click the three dots icon in the share box.**

2. **Click Poll.**

3. **Type your question in the Write Something box.**

   You can see a poll being created in Figure 10-3.

FIGURE 10-3:
You've got questions? Facebook has answers.

4. **Enter possible answers to the question in the Add Poll section.**

   Although three option fields are visible, you can add almost as many options as you can think of. To add another option field, click the +Add Option button.

5. **To choose whether people can add their own options and select multiple options, click Poll Options at the bottom of the share box and make your selections.**

   The Allow Anyone to Add Options check box controls whether people can add more answers to a poll. If you add only two options, Dog or Cat, for example, group members may be able to add Hedgehog if they want. Depending on how big your group is, this choice may or may not be significant. You can also choose whether people are limited to one selection or can choose multiple answers from the poll.

6. **Click Post.**

   The question appears in the group and in members' notifications and News Feeds. They will be able to vote, like, or comment on the poll.

Creating events

Your group may be based around an activity, so Facebook makes it easy for people to plan events for group members, whether it's a game of pick-up basketball or a birthday dinner. To create an event, follow these steps:

1. **In the share box, click the three dots icon.**

2. **Click Create Event.**

   The Create Event pop-up window appears, as shown in Figure 10-4. The fields here are the same ones you see when you plan a regular event, as detailed in Chapter 13.

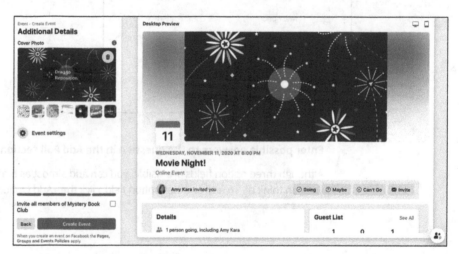

FIGURE 10-4:
Creating an event for group members.

3. **Choose to create an online or in-person event.**

   Online events can be held in Messenger rooms (see Chapter 9), or through Facebook Live (see Chapter 11), an external link, or another method. While filling out event details, you will be prompted to choose the method through which participants will join your online event. If you choose Other, be sure to provide details for how to access the event.

4. **Fill out the event details.**

   Event details include the event's name, location, date, time, and an event photo. After each step, click the Next button to proceed. On the last step in filling out the details, you see a check box with an option to invite all members of the group to the event.

5. **Click Create Event.**

   The event's page appears. Here you can adjust details and keep track of RSVPs. (We cover event maintenance in detail in Chapter 13.) As the event creator, you're automatically listed as attending. The post appears in the group's recent posts and in members' News Feeds.

6. **(Optional) Send invites.**

   Click the Invite button to see a list of members and send them an invite. Group members will receive the invite and know it was from you. This feature is helpful in large groups when members have turned off notifications and aren't checking the group daily. If you forego sending invites, all members will still be able to view the event and RSVP.

To RSVP to a group event, follow these steps:

1. **In the Recent Posts section of the group's Discussion tab, click the event's name.**

   The Event home page appears, as shown in Figure 10-5. This page shows you more information about the event, including who has already RSVP'd.

2. **Choose Going, Maybe, or Can't Go.**

   These options are big buttons towards the top of the page, below the event photo.

Using files and docs

Features that are particularly helpful to groups representing real-world projects are files and docs. These are ways to create and share files among group members. *Docs* are more like wikis in that they can be edited by all members of the group. *Files* are more like a file-sharing system that allows people to upload and retrieve files from user to user.

FIGURE 10-5:
Checking out a
group event.

Docs

Docs are sort of a cross between a wiki and a blog post. The way you create them and the way they look mimics the look and feel of a blog post, with a cover photo on top of the doc and the title and text below. You can simply publish a doc to your group or enable all members of the group to edit it as well.

To create a document that all group members can see and edit, click the three dots icon in the share box and choose Create Doc. This brings you to the doc creation page, where you can enter a title and body text. Some basic formatting options are available by clicking the paragraph icon, as shown in Figure 10-6. After you enter your text, click the Save or Publish button. *Saving* your work saves it as a draft (no other group members will be able to see it yet). *Publishing* makes the doc visible to all members of your group.

By default, all group members can edit the document. If you don't want them to do this, deselect the Allow... check box at the bottom left of the doc creation window. Assuming you leave it selected, any member of the group can view the document, comment on it as though it were any other post, and click Edit above the cover photo to change it.

You can find existing docs by clicking the Files tab and choosing the title of the document you want to look at.

Hudson Valley Wine Trail

Files

Like docs, files are shared among group members. But unlike docs, they can't be edited in Facebook. Instead, members upload, download, and reupload files they want changed. Or they simply upload files they want to share, and other group members can download the files to their own computers.

Click the three dots icon in the share box, and then click Add File. Next, click the Choose File button to browse your computer for a file that's 100 MB or less to upload. Then click Post. To download files that others have added, click the three dots icon the right of the filename and select Download from the drop-down menu.

REMEMBER

Depending on the settings of the group, you may not be able to add files without the approval of a group admin.

TIP

You might find other types of posts useful in a group setting. A well-placed Ask for Recommendations post can be a lifesaver. One of our neighborhood groups regularly posts recommendation requests for local home specialists, such as plumbers or roofers. Write a Prompt is a fun way to increase group engagement around a certain topic. The Raise Money post is a fantastic tool for charity-based groups. (We explain the ins and outs of fundraising in Chapter 12.)

Reading and commenting on posts

Things really get interesting after you create a post or see a post that someone else has created because members of the group can start talking about it. On Facebook, "talking about it" means commenting, liking, and following posts.

One unique feature of posts to Facebook Groups is that — in smaller Facebook groups, at least — you can see who has read those posts. At the bottom of each post, look for text that says Seen by <number of people>. If you hover your cursor over the number, you'll see a list of the people who saw your post. They haven't necessarily read your post in depth or interacted with it beyond a quick glance. This feature is useful when you need to know if important information has been conveyed. For example, if your choir group changes the location of its next practice, the director will have an idea of who has seen the update.

Below each post are two actions you can take:

>> **Like/React:** When you like anything on Facebook, the person who created the content is notified that you like it. Liking is an easy and quick way to say, "Good job!" when you don't have an active comment to make. Hover your cursor over the Like icon for the full menu of reactions: Like, Love, Care, Haha, Wow, Sad, Angry.

>> **Comment:** If you want to voice an opinion, click Comment below the post. You'll see a blank comment box already open below the post, just inviting you to chime in. (This action means you'll be notified about all subsequent posts.)

Depending on your notification settings, you may be following all posts in a group, only ones from friends, or none. When you comment on a post, you automatically start following that post. *Following a post* means you're notified every time there's a new comment on that post, which can be awesome if you're actively talking about something with group members or annoying if too many people are commenting. If the notifications are bothering you, you can always unfollow a post by taking these steps:

1. **Click the three dots icon in the upper-right corner of the post you want to unfollow.**

   This opens a menu of options.

2. **Click Turn Off Notifications for This Post.**

TIP

Just as you can unfollow a post you previously commented on, you can choose to follow a post you haven't commented on. If you want to read what others have to say but don't have anything to add at this time, follow the preceding steps for the post you want to follow. Instead of the Turn Off option, you will see a Turn On option. Select it to be notified every time a group member adds a new comment on that post.

Group Dynamics

Now that you know how to create, share, and join a group, it's time to look at some of the ways to manage your engagement with the group.

Controlling notifications

Sometimes, especially in larger groups, you may become overwhelmed by all the notifications. To control them, you need to get comfortable with the Notifications Settings menu, which you access by clicking the three dots icon from a group's Home page. Choose Manage Notifications from the drop-down menu to open a pop-up window with five options, as shown in Figure 10-7.

>> **All Posts:** Comment threads often become long and rambling. If you select this option, you'll know when a new post is created but you won't receive notifications whenever someone comments on that new post.

FIGURE 10-7:
Controlling
notifications.

>> **Highlights:** Select this option if you want notifications for posts that have lots of likes and comments as well as any posts made by friends.

>> **Friends' Posts:** In especially large groups, you might not be Facebook friends with everyone in the group, so a good way to filter content is to pay attention to only the posts your friends make.

>> **Off:** Some people may not want to receive any notifications from a group and instead read posts only when they choose to look at the group. Selecting Off gives you that silence while still alerting you to any group setting or privacy updates and any posts in which you're tagged.

>> **Member Request Notifications:** Select this option if you want to be informed when your Facebook friends ask to join the group. If you switch this setting to off you'll see your friends' requests to join only when you visit the group itself.

Remember to click Save to update your settings.

If your goal is to remove group posts from your News Feed but remain a member of the group, you can unfollow the group. Click the three dots icon and choose Unfollow Group from the drop-down menu. Voilà!

Searching a group

If your group is particularly active — meaning people are constantly posting content and commenting — you may have trouble finding a post. Each group has tabs below its cover photo for things such as Members, Events, Media, and Files. To the right of the tabs is a search box. Click the magnifying glass icon to search posts by keyword. For example, in a houseplant group, you might search for *Monstera care instructions*. The results might look overwhelming at first, but you can filter by Posts You've Seen, Most Recent, Posted By, Tagged Location, and Date Posted.

To locate one of your groups, click Groups in the left sidebar of the Home page or the group icon (three people) in the top bar. Then click the name of the group you want to open. Groups you visit frequently are also listed in the Shortcuts section of the left sidebar.

TIP

You can pin a frequently visited group to your Home page. Click the three dots icon from the group's Home page and choose Pin Group to ensure that the group stays at the top of your groups list in the left sidebar.

Adding friends to a group

In most groups, members can invite their friends to join at any time. This is a fast and easy way for groups to get all the right people even if the original creator isn't friends with everyone in the group. For example, you could invite a coworker to join your book club group even though that person isn't a Facebook friend with anyone else in the group.

You can add friends to a group by using Facebook or email. Simply click the blue Invite button to the right of the group name. You'll see a list of friends with a check box to the right of each name, as shown in Figure 10-8. Select the box if you want to send that friend an invite. You can also enter the name or email address of your friend in the search box up top; Facebook autocompletes as you type.

FIGURE 10-8:
Adding friends
to a group.

When you've added the friend's name or email address, click the blue Invite button. You may also see some friend suggestions under the Choose Friends box. Facebook suggests these friends based on who you interact with the most or who Facebook has determined will enjoy the group.

Creating Your Own Groups

Now that you understand how to use groups, you may find an occasion to create a group. As a group's creator, you're by default the *group administrator,* which means that you write the group's information, control its privacy settings, and keep it running smoothly. You can also promote other members of the group to administrator, thereby granting them the same privileges, so they can assist you with these responsibilities.

Here are the steps to create a group:

1. **In the left sidebar of your Home page, click Groups.**

   Your Groups page appears.

2. **In the left sidebar, click the blue Create New Group button.**

   The Create Group window appears, as shown in Figure 10-9.

FIGURE 10-9:
The Create
Group page.

3. **In the Group Name field, enter a name.**

   Choose something descriptive, so that when you add people to the group, they'll have an idea what it's about.

4. **Click the black arrow in the Choose Privacy field.**

   Groups have two privacy options:

   - *Public:* Public groups are available to the public. Anyone can request to join by clicking a Join button, although the person's acceptance may be subject to admin approval. Anyone can see all the content that the group posts. This type of group is best for a public organization that wants to make it easy for people to join and contribute.

   - *Private:* If you choose to make your group private, only members will be able to see other members and their posts. You also decide whether or not you want to keep your group visible, which means it is searchable by any

Facebook user who can request to join the group, or hidden, which means only members will be able to locate and participate in the group. Keep in mind that all group types allow members to invite their friends and give admins the option to manually approve member requests.

5. **Type the names of people you want to add to the group from the Invite Friends field.**

At this time, you can add only friends as members. Facebook tries to autocomplete your friend's name as you type. When you see the name you want, click it. You can add as many friends as you like. If you forget someone, you can always add the person later.

6. **Click Create.**

Your group's home page appears. Note the new Admin Tools menu in the left sidebar.

Adding detail to your group

Now that your group is live, you need to attend to some housekeeping items so that your group is a place where members want to spend their time.

Adding a cover photo

Facebook adds a default cover photo, but if you want something that better reflects your group, click the grey Edit button in the lower-right corner of the cover photo. From here you'll be able to choose from group photos (best for established groups that have a litany of photos to choose from), upload a photo from your computer's hard drive, choose from your Facebook photos, or choose one of Facebook's illustrations. Click to select the photo or illustration you want to use. You return to the group page, where you can click and drag the cover photo to reposition it. Click the blue Save Changes button at the top of the photo when you're done.

You can change your group's cover photo at any time by hovering your mouse cursor over the cover photo and clicking the Edit button. You can then choose where you want to select a photo from (group photos, your photos, or uploaded from your computer).

Adding a description

When people are deciding whether to join a group, they need to know about the group's purpose. One way to provide that information is to create a group description. Choose Settings from the Admin Tools menu. Then click the pencil icon next to Name and Description in the Set Up Group section. Add a Description in the box and click Save. The description lives in the About section of your group, and you can edit it from Admin Tools at any time.

Deciding a group type

All new groups are created as general groups, and most groups are in this category. *General* just means the basic group format with all the settings you'll need to manage a large or small group. However, the following groups have additional controls and features:

>> **Buy and Sell:** Members in this type of group see an additional Buy and Sell tab, where they can create sale posts and specify the currency. For example, a rare-book collecting group might want to facilitate sales of first edition finds between members. Head to Chapter 12 for the skinny on selling and buying on Facebook.

>> **Gaming:** The focus here is on setting up tournaments and linking to games.

>> **Social Learning:** This type of group can organize posts into themed units or modules. Note that you can turn on social learning units for most group types from the Add Extra Features section of the Settings page.

>> **Jobs:** Use this group type to create searchable job listings.

>> **Work:** This is a group for members who work at the same company. You'll be able to communicate with people you aren't Facebook friends with and see custom options such as the ability to request a shift switch.

>> **Parenting:** Parenting groups allow members to post anonymously and set up mentoring relationships. For members to become mentors, they simply head to the Mentorship tab, click the Become a Mentor button, and submit their qualifications or skills. Members see the list of mentors in the Mentorship tab and can click to start a conversation in Messenger, so that both parties can determine if they want to commit to ten weeks of a one-to-one mentoring exchange. You can turn on Mentorship options for most group types from the Add Extra Features section of the Settings page.

To change your group type, head to Settings from the Admin Tools menu. Then in the Add Extra Features section, click the pencil icon next to Group Type. Choose your group type and click Save.

Being a Group Administrator

If you're the creator of a group, you're automatically its *admin*, or administrator. Additionally, you can be added as an admin of someone else's group. After you have members in your group, being an admin means that you have a few extra features available to you, such as scheduling posts and pinning announcements.

You also have a suite of admin tools available to customize your group to your specific goals.

Scheduling posts

Often group admins find they want to post something at certain times. If you're the admin of your college's alumni group, you might want to post a link to sign up for the reunion as soon as it goes live. Or if your group is about sharing inspirational quotes, you might need to post an inspirational missive every morning at the same time, so group members wake up to a new quote in their News Feed. Scheduling posts allows you to create a post and choose a time in the future when it will be published.

To schedule a post, do the following:

1. **Create a post following the same steps detailed in the "Writing a post" section.**

   Simply follow all the instructions but don't press Post.

2. **Click the calendar icon to the right of the Post button.**

   A window appears for scheduling the date and time of your post.

3. **Click the Date field and select a date in the calendar that appears.**

   You can select a date up to three months in the future.

4. **Click the Time field and choose a 5-minute increment between 12am and 11pm.**

5. **Click the blue Schedule button.**

You can view all the posts you have scheduled in the Scheduled Posts section of your Admin Tools, where you also have the option to reschedule the post or post it immediately.

Pinning announcements

If you have a large group that has a lot of activity, especially a lot of people joining over time, you may find that similar posts occur on a regular basis. If you think that certain posts need top billing, admins can turn these posts into pinned announcements, which appear at the top of the Discussion and Announcement tabs before more recent posts. For example, a choir group might have the seasonal rehearsal schedule pinned to the top of the announcements for members to reference.

To make a post an announcement, follow these steps:

1. **Click the three dots icon in the upper-right corner of the post you want to change.**

2. **Click Mark as Announcement from the drop-down menu.**

   The post is now pinned to the top of the page. You may make up to 50 announcements per group.

Announcements don't have to be just for new members, and they don't have to last forever. To remove an announcement from the top of the page after it has outlived its usefulness, follow click the three dots icon on the post and choose Remove Announcement.

Managing a group

Facebook has spent the past few years improving groups and giving admins a robust set of controls. We think it's important to cover most of these controls because they will help ensure that your group members have a positive experience. With this knowledge under your belt, your group is sure to thrive and attract new members. When you're looking at your group's Home page, the Admin Tools menu (shown in Figure 10-10) appears on the left side of the screen.

FIGURE 10-10: Admin tools at your disposal.

The following Admin tools help you create a rewarding and supportive group environment:

>> **Member Requests:** Group admins often want to vet new members, simply to keep the group's intentions intact and make existing members feel comfortable with new people joining. You can learn more about prospective members here, including how long they've been a part of Facebook and whether or not a friend invited them to the group.

>> **Automatic Member Approvals:** If approving every individual member sounds tedious, you can decide to automatically approve members who meet certain requirements, such as being Facebook friends with another group member or living within 25 miles of a specified city.

>> **Membership Questions:** More and more group admins are adding a questionnaire for prospective members to fill out while they wait for their membership approval. Admins add up to three questions that can run the gamut from serious to silly. A houseplant group may ask you to name your favorite flowering plant. Only admins and moderators can see the answers.

>> **Pending Posts:** Admins may require that they approve every post before it's added for all members to see. This feature keeps disrespectful or unwanted posts off the Discussion tab.

>> **Post Topics:** This option is a way for admins to search all the posts that fall under a topic created and organized by the admin or group members. For example, if you post a chocolate chip cookie recipe to a recipe-sharing group, you can add a post topic by clicking the three dots icon in the upper-right corner of the post and typing a topic, such as desserts. Down the line, an admin for the group can search the desserts topic to find and repost the delicious recipe.

>> **Scheduled Posts:** This option contains a list of every post that the admin has previously scheduled.

>> **Activity Log:** Here you see a big-picture snapshot of the group's entire activity history, with newest activity showing first.

>> **Group Rules:** It's important to set the tone of your group, and group rules help with just that. Add up to ten rules to guide member interactions and make clear the kind of behavior that will not be tolerated. Many groups have rules regarding respect and a ban on cursing or name-calling.

>> **Member-Reported Content:** Groups run on a feeling of community, and this feeling can be difficult to maintain in larger groups where members don't always know one another. Members have the option to report group content that they think is offensive or violates group rules. The admins can then decide to remove the content. Members also have the option to report content to Facebook for review.

>> **Keyword Alerts:** Admins can set up keyword alerts to assist in flagging content for their review without waiting for a member to report it.

>> **Group Quality:** Facebook will let admins and moderators know if they've had to take action against content in the group that violated Facebook's community standards. If repeat offences occur, Facebook may turn on post approval for certain members or lower the group's search ranking.

Adjusting group settings

The final menu in your Admin Tools arsenal is Settings. These settings help set the vibe for your group. We highlight some of those settings next:

>> **Web Address:** You can create a custom URL for your group, such as www.facebook.com/groups/mysterybookclubbrooklyn/. After your group has 5,000 members, you will no longer be able to change this setting.

>> **Group Color:** You don't have to stick with the default Facebook colors. Go ahead and choose a different color to use for your group's buttons and accents.

>> **Badges:** *Badges* are little symbols that show up next to a member's name and identify the person as a certain type of member. It's up to you whether or not to keep the six default member badges turned on. We highly recommend keeping the badges turned on for Admin and Moderator so that members know who runs the show and who they should go to with any questions or concerns. Other badges are New Member, Rising Star, Visual Storyteller, and Conversation Starter (the last three reward members who create engaging posts).

>> **Rooms:** Turning on this feature allows members to interact with one another over video chat in Messenger. Room creation can be limited to admins and moderators or done by anyone in the group. Admins may also opt to include a chat window in their rooms. Room chats are saved in Messenger. (See Chapter 9 for more about rooms.)

>> **Who Can Join the Group:** Admins can limit members to those with Facebook profiles or open up membership to Pages as well. Suppose that you mostly use Facebook as your company's Page. You could join a Brooklyn Entrepreneur group that allowed Pages as members, which would make it easier for you to interact using your business persona. Learn all about Pages in Chapter 14.

>> **Apps:** Groups can add applications that provide a useful service to members or make the group run more smoothly. We discuss apps in Chapter 15, but for now, just know that you can search from a list of apps to add to your group.

REMEMBER

Click the Save button when you've made a change to your group's settings; otherwise your hard work will be lost.

TIP

Sometimes you can find additional settings on Facebook's mobile site. For example, admins have the option to add up to five tags to help surface the group to potential new members. For example, a book club might tag *bestseller* to appeal to fellow book lovers. The Tags option is part of the Basic Group Info section in the mobile Admin Tools menu.

Interpreting insights

Now that you're a wiz at managing your group's settings, it's time to gain some insight. Below Admin Tools, you'll find options for monitoring your group's data:

>> **Growth:** Set a time frame and view total members, active members, and membership requests. Members are considered active if they have engaged with group content or viewed content in the group or via their News Feed.

>> **Engagement:** See a detailed view of the number of posts, comments, and reactions for any given time frame while highlighting particular days where interaction was high.

>> **Admins and Moderators:** See what your fellow admins and moderators are up to in terms of approving and removing memberships and posts.

>> **Membership:** Get a breakdown of member demographics, including gender, age, top countries, top cities, and top contributors.

>> **Mentorship:** If you have this feature turned on for your group, you'll be able to see how many mentor and mentee pairs have formed and are currently active and how many members are seeking mentors and have yet to be paired.

Editing members

As an admin, you can remove and ban members from the group, as well as create other admins and moderators to help shoulder the burden of admin-hood.

To edit members, follow these steps:

1. **Click the Members tab on your group's home page.**

The Members section of the group appears, the bulk of which is taken up by images of group members. If you are an admin, a three dots icon appears to the right of each group member's name.

2. **Click the three dots icon next to the member's name.**

The menu shown in Figure 10-11 appears.

3. **Make your selection:**

- *Make a member an admin or a moderator.* Making someone an admin means the person will have all the same powers as you to add new admins, edit the group's privacy (and other) settings, and so on. Moderators have the same abilities to review requests to join, schedule posts, and remove members but cannot add more admins or edit the group's settings.

- *Turn on post approval.* If certain members are troublesome and their posts are being reported by other members, it may be worth turning on post approval so that every post they make is accepted by an admin before appearing to the group.

- *Mute the member.* Another option for problematic members is to mute them for up to 28 days, during which time they'll be able to see the group but not post or comment. It's fairly obvious to members when they've been muted, so we suggest turning on post approval as a first step.

- *Remove or block the member.* If the members continue their bad behavior, it's time to consider permanent removal from the group or even blocking them from ever finding the group again. If you choose to remove someone from the group, a window appears with options for deleting that member's posts, as well as blocking them permanently and removing any pending invites they've sent to others.

  When someone is an admin or a moderator, you can remove their status by choosing Remove as Admin or Remove as Moderator from this same menu.

We hope you never need to use these options, but unfortunately that's just not always the case.

Reporting offensive groups and posts

If you stumble upon an offensive group in your day-to-day scrolling, you should report it to Facebook so that the company can take appropriate action. To report a group, follow these steps:

1. **Click the three dots icon under the group's cover photo and select Report group.**

   A form appears in a pop-up window.

2. **Choose a reason for the report.**

3. **Click Next to send the report.**

   Facebook may suggest additional steps you can take, such as leaving the group.

4. **Click Done to close the report pop-up window.**

Facebook attempts to remove groups that

» Contain pornographic material or inappropriate nudity

» Attack an individual or a group

» Advocate violence

» Serve as advertisements or are otherwise deemed to be spam by Facebook

REMEMBER

Many groups on Facebook take strong stands on controversial issues, such as abortion or gun control. In an effort to remain neutral and promote debate, Facebook won't remove a group because you disagree with its statements. However, you can report hate speech and false news when you encounter it because these violate Facebook's community standards.

To report offensive content within a group, such as a post written by a group member, follow these steps:

1. **Click the three dots icon at the upper-right corner of the post.**

   A menu of options for that post appears.

2. **Choose Report Post to Group Admins, Find Support, or Report Post.**

   Reports to admins won't notify Facebook, but reports to Facebook will notify the admins. You will need to verify that you want to report a post, and reports to Facebook will request more information about why you're reporting something.

Depending on the nature of your group, reporting content to your admins might have the same effect as reporting it to Facebook. However, admins may be either more lenient or harsher than Facebook. Some Facebook groups have strict "be kind" policies, so a post reported to Facebook might be permitted to remain, whereas the admins or moderators may deem it too mean for the group. As with most reporting questions, we advocate for reporting the things that concern you.

4
Getting the Most from Facebook

IN THIS PART . . .

Creating photo albums and sharing photos and videos

Fundraising for a good cause

Creating and managing events

Creating and interacting with Pages

Using games, websites, and apps with Facebook

Chapter **11**

Filling Facebook with Photos and Videos

acebook Photos along with Facebook-owned Instagram are the top photo-sharing platforms on the web. And the fact that *all* your friends are likely on Facebook and using Photos makes it a one-stop shop for uploading and tracking all the photos of you, all the photos you've taken, and all the photos of your friends.

Additionally, Facebook Photos allows you to add and share videos. Although a little less common, videos are pretty similar to photos. If you let them languish on your hard drive or on your mobile phone, nobody gets to enjoy them. Nobody gets to tell you how cute your baby is. No one can tell you that they like your wedding video. But when you share photos and videos, they can become more cherished and even more valuable as keepsakes.

Viewing Photos from Friends

Just by opening Facebook and looking at News Feed, you'll find yourself looking at lots of people's photos. You'll see photos in a few different ways: in your News Feed, in the photo viewer, and in an album format. However, in a world of ever-diminishing attention spans, you're more likely to come across single photos posted in the moment versus albums filled with hundreds of photos.

Photos in News Feed

Figure 11-1 shows an example of how a single photo appears in News Feed. The photo takes up most of the screen. Across the top is the name of the person who posted it and any description they wrote about it. You can also see info about when the photo was added and possibly where it was added (for example, Indianapolis, Indiana). Below the photo are icons to Like and Comment on the photo and sometimes an icon to Share the photo. Above these icons is the count of how many likes and reactions the photo has already received. Below the buttons are any comments people have made. You'll likely see a blank comment box here as well, waiting for you to add your two cents.

FIGURE 11-1:
Regarding a friend's photo in News Feed.

Clicking the photo expands the photo viewer, which is covered in the following section.

Figure 11-2 shows an example of multiple photos in News Feed. It's similar to the single photo, but previews additional photos from the collection. This is not a photo album post, although these photos will be sorted into automatic albums

that Facebook creates. You can identify a photo album post in your News Feed because photo albums have titles and clicking the title from the post brings you to an album view, which we go over later in the chapter.

Photo viewer

The photo viewer is an overlay that allows you to quickly browse photos and leave likes and comments. Clicking a small version of a photo almost anywhere on Facebook expands the photo viewer and fades the rest of the screen to black, as shown in Figure 11-3. The photo takes up the left side of the page and comments, likes, and info about the photo appear on the right side of the page.

FIGURE 11-3:
The photo viewer.

If the photo is one of many, you'll see an arrow on either side of the photo that allows you to navigate through the collection. Using your computer's forward and back arrows also does the trick. In addition to the arrows, you'll find other actions you can take on the photo itself:

>> **Tag photo:** Clicking the tag icon allows you to add *tags,* or labels, for those in the photo. We cover tags in the upcoming section, "Editing and Tagging Photos."

>> **Options:** Next to the poster's name is the three dots icon. Clicking it reveals a menu of options for things such as downloading the photo.

>> **Share:** Clicking the Share icon lets you post the photo to your own timeline, a friend's timeline, a group, a messenger chat, or a Page. Privacy settings determine whether the share feature is available for a photo.

>> **Like:** As with the Like icon that appears under a photo in News Feed, selecting this option lets the person who added the photo know that you like the photo.

REMEMBER

When we mention liking something, we are referring to all the reaction options Facebook provides. There are times when a like isn't specific enough; perhaps you prefer to love your friend's engagement photo or want to express astonishment at a photo of breathtaking views.

The album view

The *album view* is the grid of thumbnail photos that you see when you click the name of an album. Most screens can fit about 8 to 12 photos in this view, and as you scroll down the page, more photos appear until you reach the end of the album. Sometimes if people add a really large album, you may want to skim the album view to identify the parts of the album that interest you. Clicking any photo brings up the photo viewer.

To the right of the photo grid are the name of the album, any general info your friend has added about the album, and who has liked the album or commented on it. Figure 11-4 shows a sample album view.

FIGURE 11-4:
An album view of photos.

TIP

Commenting on an album is different than commenting on a single photo. Leave a comment on an album when you want to express something about the collection: "Looks like a great trip!" "He's growing up so fast!" Comment on a single photo when you have something to say about that photo in particular: "Did you use a fish-eye lens to get this shot?" or "OMG I was at this exact spot a year ago!"

Viewing photos on your mobile device

Chances are that if you have a smartphone or a tablet computer such as an iPad, you'll wind up looking at photos using the Facebook app. Looking at photos on these devices isn't too different from looking at them on a computer screen. Tapping a photo in News Feed expands the photo and fades the rest of the screen to black. At the top of the screen are icons for closing the photo and returning to News Feed (X), tagging the photo (tag icon), and more options (. . . icon). At the bottom of the screen are icons you can tap to like, comment on, or share the photo if privacy allows. Tapping the count of likes or comments expands a screen where you can scroll through the comments people have made on the photo.

When someone has added multiple photos in one post, you see a preview of those photos in your mobile News Feed. When you tap any of the photos, you see a column view, where you can scroll up and down to browse through all the photos that have been added. Tap the back arrow in the upper-left corner to return to News Feed. Tap any photo to view it in the photo viewer, and then swipe left and right to navigate through the photos. The two-finger method of zooming in and out also works on Facebook Photos.

Viewing tagged photos and videos of yourself

When we say *photos and videos of yourself*, we're referring to photos and videos in which you're *tagged*. Tags are ways of marking who is in a photo — the online equivalent of writing the names of everyone appearing on the back of a photo print. Tags are part of what make Facebook Photos so useful. Even if you don't add lots of photos, other people can add photos of you. Photos you've been tagged in might be scattered across your friends' timelines, so Facebook collects all these photos in the Photos tab below your cover photo.

The Photos tab defaults to showing Photos of You. You can also view photos you've added (Your Photos) or albums you've added (Albums). The Photos of You section shows the most recently tagged photos at the top of the page. As you scroll down, you see older and older photos of yourself. This is a great place to take a trip down memory lane and also to familiarize yourself with all the photos of you that are out there.

If you've been tagged in a photo and you don't like that tag, you can always remove the tag by clicking the pencil icon in the top-right corner of the photo and choosing Remove Tag from the drop-down menu. If you're looking at the photo in the photo viewer, you can also click the tag icon, which shows you a list of tags on that photo. Only your tag will have an X next to it; you can click it to remove your tag. When you remove a tag, the photo is no longer linked to your timeline and will cease to appear in the Photos of You section.

If you don't want a photo or video on Facebook at all, even after you've removed the tag, get in touch with your friend and ask her to remove it. If you think it's offensive or abusive, you can also report the photo and ask Facebook to remove it.

Adding Photos to Facebook

Facebook is a great place to keep your photos and videos because it's where most of your friends will be able to see them. Whether that's a single photo you snapped on your phone or a big album detailing the latest family road trip, photos are most fun when you can share them and talk about them with your friends.

As we mentioned, Facebook distinguishes between uploading photos and creating a photo album. Albums are often created to document a particular event or period of time, whereas uploads happen on an ongoing basis. Because photo uploads tend to happen more frequently, we go over all the ways of uploading photos before delving into album creation.

Uploading photos

If you have a few photos you want to quickly share, follow these steps to get them out to your friends:

1. **Click Photo/Video at the bottom of the share box on your Home page.**

A window appears allowing you to browse your computer's hard drive and select the photo you want (see Figure 11-5).

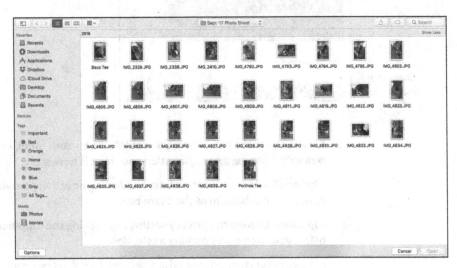

FIGURE 11-5:
Choose your
favorite photos.

2. **Click the photo(s) you want to share.**

3. **Click Open or Choose (the wording depends on your browser and operating system).**

   You return to Facebook, where you see a preview of the post including any photos you've added (shown in Figure 11-6).

WARNING

Sometimes Facebook takes a while to add your photo, in which case you'll see a progress bar or circle in lieu of the photo. You won't be able to post until the photo has finished uploading.

FIGURE 11-6:
Share your
photos with
your friends.

4. **Click in the share box (where you see What's on Your Mind, <*Your Name*>?) and type any explanation you think is necessary.**

5. **(Optional) Add tags, location info, and feeling or activity details from the options at the bottom of the share box.**

6. **(Optional) Change the privacy settings by clicking the drop-down menu below your name and making a selection.**

   If you've never changed your privacy settings, by default everyone on Facebook can see your photos if they navigate to your timeline. We usually like sharing our photos with friends. Of course, you can always choose custom groups of people who can and cannot see the photo. And you can alter the privacy setting on a post-by-post basis if some photos are more private than others. Check out Chapter 6 to become a privacy settings whiz.

7. Click Post.

You have officially shared the photo to Facebook. People can see the photo on your timeline and in their News Feeds (provided they're allowed to by your privacy settings). By default, the photo is added to an album called Timeline Photos, which is a collection of all the photos you've ever added individually.

Editing photos as you add them

Uploading photos is meant to be a no-brainer. In essence, you click the photos you want to share and then you click post. Voilà, shared photos. However, Facebook offers a bunch of cool photo-editing options that you can choose to use as you add your photos.

WARNING

This is one of those places where what you see on-screen might be different than what's pictured in the figures in this book. Don't worry, the editing options are the same, but some locations (left versus right, top versus bottom) might be different.

To get to these options, hover your mouse cursor over any of the previews of the photos you have chosen to add (do this *before* you click Post). When you hover the cursor over the photo, three buttons appear. Click X to remove the selected photo from the post. Click Add Photos/Videos if the post could do with more. Finally, click Edit or Edit All to open a photo viewer with editing options (shown in Figure 11-7):

» **Crop:** Cropping involves changing the borders of the photo. After you click Crop, guidelines appear that you can move to crop the photo. Click and drag the guidelines from one of the photo's corners until your photo is perfectly cropped. Be sure to click the Save button at the bottom of the Edit Photo window.

» **Rotate:** No need to make people tilt their heads at their computer screens to see your photo properly. Click Rotate to rotate the photo 90 degrees. Click it as many times as you need to get to the proper orientation.

» **Tag Photo:** You can tag a person in a photo by clicking her face and then typing her name in the text box that opens. Facebook autocompletes as you type; click the person's name when you see it. Tagging friends will notify them that they've been tagged in a photo and may allow their friends to view that photo as well.

» **Alternative Text:** Alternative text is intended for people who are blind or visually impaired. Most alternative text includes a description of the image, such as "Amy is wearing a gold lamé dress and kicking up her left leg." Add your own text in the box that says Custom Alt Text or choose Facebook's autogenerated text. The text is added to your image for those who need it, making your photo more accessible to everyone.

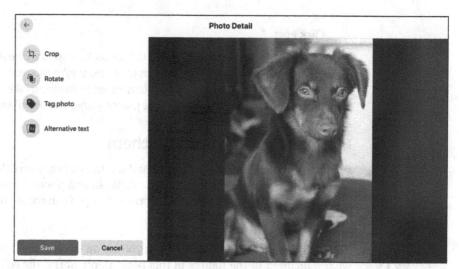

FIGURE 11-7:
Photo editing
options.

Make your changes permanent by clicking Save at the bottom of the Edit Photo window.

Adding photos from your iPhone

Many photos you see on Facebook are added when people are nowhere near a computer. Instead, they're photos of things that happen while out and about. Things that are beautiful (spring blossoms!), or strange (how did this person lose only one high heel?), or emblematic of your day (another cute photo of the dog).

If you add the Facebook app to most smartphones, you can send photos from your phone right to Facebook. We go over how to add a photo from the share box in the iPhone app in Chapter 7. You can also add a photo from your iPhone's camera roll.

To add a photo from your iPhone's camera roll, follow these steps:

1. **From your photo gallery, tap the photo(s) you want to share; then tap the Send icon (box with an arrow) at the bottom left of the screen.**

   A menu of options appears. If you previously installed the Facebook app, you should see the Facebook icon among the options.

2. **Tap the Facebook icon.**

   Choose to share the photo to your News Feed or to a story. We chose News Feed in this example. The share box appears, as shown in Figure 11-8.

3. **(Optional) Click Say Something about This Photo and then use the keyboard to type any explanation the photo needs.**

4. **(Optional) Tap the privacy icon (Friends in Figure 11-8) to edit who can see that photo.**

   Remember, by default, the audience you shared your last post with will be the people who can see this post. Your options on mobile Facebook are Public, Friends, or Only Me.

5. **(Optional) Tap the Album button to add the photo to an existing album on Facebook.**

   By default, Facebook adds your photo to a Mobile Uploads album.

6. **(Optional) Tap any of the icons at the bottom of the photo to add feeling/activity info, tags, or location info, respectively.**

7. **(Optional) Tap News Feed on the top bar to share the photo to a friend's timeline or a group instead.**

   You'll need to specify which friend or which group. Sharing your photo this way will not add it to your timeline.

8. **Tap Post in the upper-right corner.**

   If you shared to News Feed, the photo is added to your timeline as part of the Mobile Uploads album, and it may appear in your friends' News Feeds.

FIGURE 11-8:
Sharing is caring.

Editing photos on the mobile app

If you upload photos using the Facebook app on your smartphone or tablet, you'll find there are more options for editing beyond those available on your computer (see Figure 11-9). We go over a few of the most popular options here. After you add a photo to the share box, click the Edit button in the upper-left corner of the photo to do the following (and more):

>> **Add effects:** Add effects by clicking the icon in the lower-left corner of the photo. Effects run the gamut from changing your photo's color tone (how about sunset colors or maybe an elegant black and white), adding details such as neon beams or stars, or altering the style of the photo to something resembling pop art or a sketched drawing. Effects are plentiful and can bring some much-needed oomph to an otherwise pleasant but unmemorable photo.

>> **Add stickers, GIFs, and emojis:** Stickers are pretty much what they sound like: digital drawings you can stick on top of your photos. Click the icon that looks like a peeling smiley face to browse through the sticker menu for the sticker you want. There are also menus for GIFs (short moving images that play on a loop) and emojis. When you click the sticker you want, it appears in the center of your photo. Click and drag it to the desired location. Click again to rotate the sticker and change its size (drag your pointer and middle fingers apart on the screen to make it bigger, or in to make it smaller). To delete the sticker, drag it toward the bottom of the screen and drop it in the trash can icon that appears. You can add as many stickers, GIFs, and emojis as you want to a photo.

>> **Add text:** Unlike alternative text, which is meant to describe the contents of a photo, this text option appears on top of a photo and is a way of editing the photo itself to say something. Click the Aa icon to open a text box and then type to add your text. You can change the color of the text using the color palette on the bottom-left. Choose one of the color dots that appear and watch as the text changes to that color. Use the multiple lines icon to decide whether your text should appear in the right, left, or center of the text box. Click the Headline button to choose a font for your text.

FIGURE 11-9:
Get ready to unleash your inner Picasso.

Click and drag the text box to move it to where you want it to appear on the photo. Drag it out to make it bigger, in to make it smaller, and up or down to spin the text on its axis. To delete the text, drag it toward the bottom of the screen and put it in the trash can icon that appears. When you're happy with what the text says and how it appears on the photo, click another portion of the photo to remove the text box, leaving just the text behind. Click the text again to bring back the text box and the relevant editing options. You can add more than one text box to a photo.

>> **Add lines:** Click the squiggly line icon in the top-right, choose your medium from the circles at the bottom of the screen, and begin dragging your finger over the photo where you'd like to add one or more lines. You can always click Undo in the upper-left corner to remove the lines and return to your original photo.

WHAT'S THE DEAL WITH INSTAGRAM?

As you scroll through your News Feed, you may see photo posts that have the word *Instagram* next to their timestamp information. At this point, you've probably heard of Instagram and wondered if it was for you. Aren't you already putting all of your photos on Facebook? Do you need another website to keep track of?

Instagram is owned by Facebook but provides a more focused service to its over 500 million daily users. The site is only for photos and short video posts (as of the time of this writing). Many people use both websites but favor Instagram for their photo posts, which they can always share to their Facebook timeline at the same time they post to their Instagram account. In fact, Facebook creates an automatic album for all your photos that originated on Instagram.

If you tend to enjoy visual posts in your News Feed, we recommend signing up for an Instagram account as well. You can follow your Facebook friends on Instagram to see their posts, but the relationship does not need to be reciprocal. You can also follow celebrities, brands, and strangers to see what they post. For example, Amy adopted one of her dogs after she saw a photo of him on an animal rescue's Instagram feed and reached out to the organization.

Creating an album

Now we return to Facebook Photos on your computer. Whereas a single photo can share a moment, an album can truly tell a story and spark conversations with your friends. To create an album, follow these steps:

1. **From your timeline, click the Photos tab.**

2. **Click the Albums tab and then the Create Album button.**

   The opens the Create Album page, shown in Figure 11-10, which looks similar to the page for creating events, groups, and Pages.

3. **Choose who can see the album by using the Privacy menu.**

   The Privacy menu, which is the grey button under Create Album, reflects the privacy setting from the last time you posted something. If you last posted something publicly, for example, the Privacy menu displays the globe icon and Public. Your options are Public, Friends, Only Me, Friends Except, Specific Friends, or a Custom set of people.

FIGURE 11-10:
Create your
album here.

4. Edit the album info and options.

The left side of the screen contains all relevant album info:

- *Album Name:* Enter your album title here. Usually something descriptive does the trick: "Maui in December" or "Benjamin's First Birthday."

- *Description:* Click here to add context to your album. You can talk about why you took these photos, or anything else you think people might want to know about your album: "Snaps from Jill's Bachelorette!"

- *Add Contributors:* Speaking of that bachelorette party, photo albums often center on an event where many people took photos. Type your friends' names and add them to this box to grant them the ability to upload their own photos to the album you're creating. The names of anyone you've tagged in a photo from the album will automatically appear here, but you can remove them as contributors by clicking the X next to their names.

- *Sort By:* Click to open this drop-down menu, which gives you the option to sort photos by date or by dragging and dropping the photos into your preferred order. The default option is drag and drop.

- *Use Date from Photos:* Click to automatically add date and time information to all your photos at once.

5. Click Upload Photos or Videos and select your photos and videos.

This opens your computer's hard drive interface where you can search for and select the photos and videos you want to include in the album. You can select multiple photos at once by pressing the Shift or Command key, clicking the files you want, and then clicking Open or Choose.

TIP

If you use a program such as iPhoto to organize your photos, create an album there first. Then navigate to it and select all those photos to add to Facebook. You'll save yourself time trying to figure out whether you want to use IMG0234 or IMG0235.

6. **(Optional) After your photo previews upload, add descriptions to individual photos.**

The preview of each photo has a space below it. Click that space to add a caption or description for each photo.

Each photo also has three icons. The location icon lets everyone know where the photo was taken, and the clock icon lets everyone know the date and time the photo was taken. The tag icon lets everyone know who's in the photo. You don't have to tag friends in your album, but we highly recommend doing so because it allows your friends to learn about your photos more quickly and share in discussing them with you. Check out the tagging box in Figure 11-11.

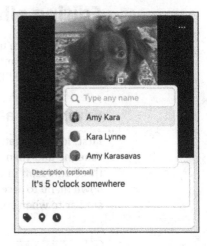
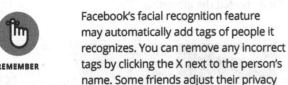

FIGURE 11-11:
Who is it?

REMEMBER

Facebook's facial recognition feature may automatically add tags of people it recognizes. You can remove any incorrect tags by clicking the X next to the person's name. Some friends adjust their privacy settings so that Facebook can't automatically recognize their face, and these friends will need to be manually tagged in each photo.

7. **Decide how to sort your photos.**

By default, your album is organized by when you uploaded the photo. You can order it by date (newest or oldest first), or you can drag and drop the photos in whichever order you think works best.

TIP

Click the three dots icon in the top-right corner of any photo to rotate the photo.

8. **Click Post.**

Whew! That was a bit of a marathon. If you need a break or a drink of water, feel free to indulge. Then, when you're ready, continue with the "Editing and Tagging Photos" section to find out how to edit your album and the photos in it.

Editing and Tagging Photos

After uploading a photo, you can still make changes to the way it appears on Facebook. If you added an entire album, you may want to add more photos or rearrange the order of the pictures. For any photo you added from a phone or just quickly from your computer, you may want to add tags, a date, or location information. Doing all this is relatively easy using the following common editing tasks.

Editing albums

Editing an album usually consists of editing the album's information or settings. You may also want to edit specific photos in an album. For those types of edits, hop down to the "Editing a photo" section.

Editing the album name, description, and privacy

First, let's navigate to the Edit Album view by choosing the Photos tab from your timeline and then clicking Albums. Locate the album you want to edit, click the three dots icon in the upper-right corner, and select Edit Album from the dropdown menu to access the Edit Album screen (see Figure 11-12). The interface looks similar to when you first created the album.

FIGURE 11-12:
Editing your album's info.

To change the album's privacy, click the grey button below Edit Album on the left side of the screen. Select your preferred privacy option from the pop-up menu. To change the album name or description, simply update the text in the Album Name or Description text fields, respectively. Remember to click the blue Save button to keep your changes.

You can take other actions from the Edit Album view. We describe some of the most common next.

Reordering photos in the album

Chances are that if you added your photos in bulk, they don't appear exactly in the right order. And it's confusing when the photos of the sunset appear first, and the photos of your awesome day of adventure come afterward. To reorder photos from the Edit Album view, follow these steps:

1. **While holding the mouse button down, drag the photo thumbnail to its correct place in the album and then release the button.**

   The other photos shift positions as you move your chosen photo.

2. **Repeat Step 1 with the next photo until your entire album is organized correctly.**

Facebook automatically saves the new order of your album, but it never hurts to click the Save button anyway.

Adding more photos

After you create a photo album, you can add more photos from the Edit Album view. To add more photos, follow these steps:

1. **Click the Upload Photos or Videos button.**

   An interface appears for exploring your computer's hard drive.

2. **Select the photos you want to add.**

3. **Click Open or Choose.**

   The upload process completes and the Add to Album page appears. Here you can tag your photos, add captions, and rotate them as needed.

4. **Click the Post button to add the photos to the album.**

Editing a photo

In addition to the actions you can take on an entire album, you can also take actions on individual photos in an album from the Edit Album screen.

Adding a tag to an individual photo

If you skipped adding tags earlier, you can always add your tags to individual photos. Just do the following:

1. **Click your friend's face.**

   The tagging box appears (refer to Figure 11-11).

2. **In the text box, enter the name of the friend you want to tag.**

   Facebook tries to autocomplete your friend's name as you type.

3. **Repeat Steps 1 and 2 until everyone in the photo is tagged.**

   Or stop after you tag a few people. You can always come back to this later.

Rotating a photo

Lots of times, photos wind up sideways. It's a result of turning your camera to take a vertical shot as opposed to a horizontal one. You don't have to settle for this:

1. **Click the three dots icon in the top-right corner of the photo.**

2. **Choose Rotate Left or Rotate Right from the menu that appears.**

3. **Keep clicking Rotate Left or Rotate Right until your photo is at the correct orientation.**

Adding or changing a description or date for an individual photo

Just as you can add a description to the album as a whole, you can add descriptions or captions to individual photos:

1. **In the box below the photo, type your description.**

2. **(Optional) Add any tags to indicate who was with you when the photo was taken.**

3. **(Optional) Add location information about where the photo was taken.**

   Facebook tries to autocomplete your location information as you type.

4. **Use the date selector to change the date when the photo was taken.**

5. **When you're done, just click Post.**

   You don't need to save your changes.

Album covers

From the Edit Album view, Facebook lets you designate a photo to act as your album cover. The cover is the first photo people see when they click on your album. To choose an album cover:

1. Click the three dots icon in the upper-right corner of the photo.

2. Choose Make Album Cover.

3. Confirm your change in the small window that pops up.

Moving photos

Sometimes you realize a photo you added to an album is better suited to a different album. There's a quick and easy way to transfer that photo, also from the Edit Album view:

1. Click the three dots icon in the upper-right corner of the photo.

2. Choose Move to Another Album.

A pop-up window appears with a drop-down menu of all your photo albums.

3. Choose the album you want to move the photo to.

4. Click Move Photo.

Deleting a photo

Maybe you realized that all 20 group shots from the family BBQ don't need to go in the album, or that one photo has a lot of closed eyes. From the Edit Album view, you can remove photos entirely from Facebook:

1. Click the three dots icon in the upper right-corner of the photo.

2. Choose Delete Photo.

A pop-up window appears asking if you're sure and letting you know that deleting the photo will also delete any posts associated with adding it to the album.

3. Click Delete.

A small pop-up window in the lower-left corner confirms the photo's removal. You return to the album view, now with one less photo.

Deleting an album

If you decide to delete the entire album, head to your Albums tab and locate the album you want to delete. Then click the three dots icon in the top-right corner of the album and choose Delete Album from the drop-down menu. You're prompted to confirm the deletion. Click the blue Delete Album button to remove the album once and for all.

If you delete your photo album, all the photos in it will be gone forever, so make sure you want to get rid of it completely before you delete it.

Automatic albums

Most of the time when you're creating a photo album, you decide what to title it and which photos go into it. There are a few exceptions to this rule. Facebook assembles certain types of photos into albums on your behalf. Most importantly, every time you change your profile picture or cover photo, Facebook adds it to the Profile Pictures or Cover Photos albums, respectively. Facebook creates albums of all your profile pictures and cover photos automatically.

You can access this album by clicking your current profile picture or cover photo from your timeline. This takes you to the photo viewer, where you can click through your historical record of pictures. Even though your current profile and cover photos are always public, you can edit past photos to change the privacy setting; add tags, a location, or a description; or delete the photo. To start, simply click the three dots icon at the top-right corner of the page. Make sure to click Save or Save Changes where applicable.

You can turn any photo from this album back into your profile picture by selecting Make Profile Picture from the three dots menu.

Similarly, photos that you add to your timeline are collected in the Timeline Photos album. Photos that you add from your phone are added to a Mobile Uploads album. Videos are collected into a Videos album. You get the idea.

Working with Video

Too often, videos wither on hard drives, or cameras, or mobile phones. The files are big, and they can be difficult to share or email. Facebook seeks to make sharing videos easier. So film away and let everyone see what you've been up to.

Viewing videos

You'll mostly encounter videos in your News Feed, with an enormous play button in the center. If you pause in your scrolling to look at the video, it will begin playing but without sound (see Figure 11-13). If you aren't interested in the video, keep scrolling. If you are, hover your mouse cursor over the bottom of the video and click the muted sound icon (a megaphone with a white X). This will turn on the sound for the video.

FIGURE 11-13:
A video on
Facebook.

Videos play inline in your News Feed. While the video is playing, you can hover your mouse cursor over it to see the progress bar, change the volume, switch the video quality, or expand it to full screen.

TIP

Do you wish that you could keep scrolling and watch a video at the same time? You can! Just click the Continue Watching while You Use Facebook icon (shown in the margin) to move the video to the lower right of your Home page and out of the News Feed. The video stays put as you continue scrolling.

WARNING

If you like (or follow) a lot of Pages, your News Feed may be filled with videos. For example, Amy follows a Page that posts nothing but videos of unlikely animal friends all day long. Some of these videos play for 15 or 30 seconds before being interrupted by a video advertisement. Just kick back and enjoy the brief ad. The video you're watching will return when the ad finishes.

Adding a video from your computer

Uploading a video to Facebook includes going out into the world, recording something, and moving it from your camera to your computer. Now, to upload a video to Facebook, follow these steps:

1. **Choose Photo/Video in the share box at the top of your Home page or timeline.**

    A window appears to allow you to navigate your computer's hard drive.

2. **Select a video file from your computer.**

    This brings you back to Facebook, where your video appears in your post preview.

3. **(Optional) In the What's on Your Mind, *<Your Name>?* area, type an explanation or a comment**

4. **(Optional) Select who can see this video using the Privacy menu.**

    As usual, your options are Public, Friends, Friends Except, Specific Friends, Only Me, and Custom.

5. **(Optional) Use the icons at the bottom of the share box to add tags, feeling/activity or location information.**

6. **Click Post.**

    Videos files are large, so if you are notified that your video is processing, it means you need to wait a little while until the video is ready. You can use Facebook in the meantime and wait for a notification that your video has uploaded.

Adding a video from the Facebook app

Much like photos, many of the videos you want to share most are ones you take when you're out and about: watching dolphins leap from Caribbean waters, your dog chasing a tennis ball, the bride and groom cutting the cake. More and more often, you may find yourself using your phone to record these videos.

You could move the video from your phone to your computer and then add it to Facebook, or you could skip the middleman and share it directly from your phone using the Facebook app:

1. **Tap the Photo option in the share box at the top of your mobile News Feed.**

   Your phone's camera roll appears.

2. **Tap the video you want to add.**

3. **Tap Done in the upper-right corner.**

   You see a preview of your video in the mobile share box. You can play it here to make sure you want to share it.

4. **(Optional) Tap the Say Something space above the video, and add a caption or description of the video.**

5. **(Optional) Use the icons on the bottom of the share box to add tags, feeling or activity, or location info to your post.**

6. **(Optional) Click Edit in the top-left corner of the video to add effects, trim the clip, or add stickers and text.**

   It's a good idea to trim the video of any extra footage and get straight to the action. Click the scissors icon from the Edit page and slide left or right to shave seconds from the beginning or end of the video.

7. **Determine the video's privacy from the drop-down menu under your name.**

   Remember, by default the video will be shared with the same group of people you last shared a post with.

8. **Tap Post in the upper-right corner of the screen.**

   The video is added to Facebook. Your friends will be able to see it on your timeline and in their News Feeds (depending on your privacy settings, of course).

Live video

There may be instances when something you're witnessing should be shared with your friends in real-time. Maybe a deer is on your lawn or you're caught in a particularly nasty storm. This is where live video comes into play. To create a live video, click the Live Video button at the bottom of the share box and follow the prompts to give Facebook access to the camera on your computer or mobile phone. The mobile version is more intuitive — you simply click the Start Live Video button.

The desktop version asks you to set up your stream. Unless you're aiming for high production value, which requires more in the way of technical setup, simply title your video, add a brief description, choose Use Camera and click the big blue Go Live button. You know your video is live because of the End Live Video red button and the word *Live* in the upper-left corner of the video.

When you end a live video, you'll see a menu of options, as shown in Figure 11-14. You can trim the video, create a clip from the video, or delete the video. If your friends miss the live version, they'll be able to catch the video later on your timeline or in their News Feeds.

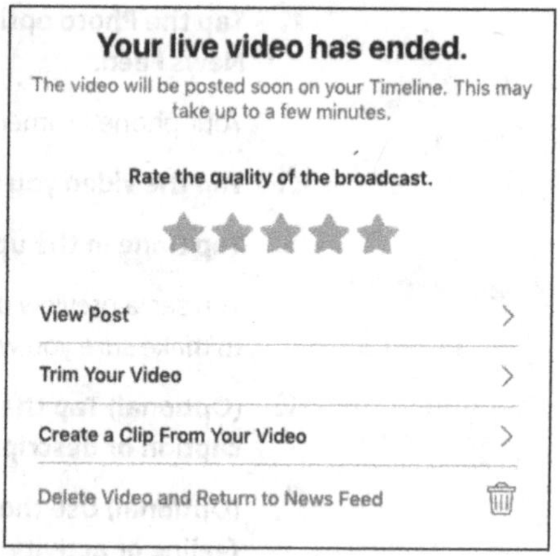

Your live video has ended.

The video will be posted soon on your Timeline. This may take up to a few minutes.

Rate the quality of the broadcast.

⭐ ⭐ ⭐ ⭐ ⭐

View Post ⟩

Trim Your Video ⟩

Create a Clip From Your Video ⟩

Delete Video and Return to News Feed 🗑

FIGURE 11-14:
Live it up on Live video.

Discovering Privacy

While privacy is covered in detail in Chapter 6, it's worth going over a few settings again here so that you understand what you're choosing to show or not show people.

Photo and video privacy

Each time you create an album, post a photo, or add a video to Facebook, you can use the Privacy menu to select who can see it. The options are as follows:

» **Public:** This setting means that anyone can see the post. It doesn't necessarily mean that everyone *will* see the post though. Facebook doesn't generally display your content to people who aren't your friends. But if, for example, people you didn't know searched for you and went to your timeline, they would be able to see that post.

» **Friends:** Only confirmed friends can see the post when you have this setting.

- » **Friends Except:** List the friends or friends lists who you don't want to see the post. Be careful with this setting. If mutual friends see the post, they may blow your cover.

- » **Specific Friends:** Show the post only to specific friends or friends lists who you know will be the most interested in the content.

- » **Only Me:** Only you will be able to see that photo or video. You might use this setting if you want to test a post before you commit to publishing it to the world at large.

- » **Custom:** Custom privacy settings can be as closed or as open as you want. For example, you may decide that you want to share an album only with the people who were at a particular event.

TIP

Another way to control who sees an album or video is to share it using Facebook Groups. So, for example, a video of your kids playing might be of interest only to people in your family. If you have a group for your family, you can share it from the share box in the group, and then only people in the group will be able to see it. Learn more about Groups in Chapter 10.

Privacy settings for photos and videos of yourself

The beauty of creating albums on Facebook is that it builds a giant cross-listed spreadsheet of information about your photos — who is in which photo, where those photos were taken, and so on. You're cross-listed by name in photos that you own and in photos that you don't own. However, if you want more control over those name tags and who can see them, click the down arrow (Account menu) in the top bar and then click Settings and Privacy. Next, on the left side of the page, click the Profile and Tagging section.

The following settings can help you further control who can see photos and videos of you:

- » **Who Can See Posts You've Been Tagged in on Your Profile?** Regardless of whether or not you review tags before they appear on your timeline, you can control who can see posts where you've been tagged. Click the Edit link to the right of this setting and make your selection from the drop-down menu. So, for example, if you set this option to Friends and you get tagged in a photo, even though it has been added to your timeline, a non-friend visiting your timeline would not see the photo displayed there.

>> **When You're Tagged in a Post, Who Do You Want to Add to the Audience If They Can't Already See It?** Another way to limit who can see that you've been tagged in a post is to change this setting from Friends to Only Me or Custom. By default, if you tag Carolyn in a photo, her friend Stephanie, whom you're not friends with, will be able to see the photo. If Carolyn changes this setting to Only Me, then Stephanie cannot see the photo you've tagged Carolyn in.

>> **Review Posts Friends Tag You in before They Appear on Your Profile?** Turning this option from off to on means you get to make sure you want to be tagged in photos (and other posts) before anyone can see that you've been tagged. Suppose that you tag your friend Carolyn in a photo:

- If this option is *off,* as soon as you tag her, the photo is added to her timeline and (usually) her friends will be able to see that she's been tagged in their News Feeds.

- If this setting is *on,* she has to approve the tag before it appears on her timeline and in her friends' News Feeds. The photo itself still gets shared and is visible on your timeline and in your friends' News Feeds.

Chapter **12**

Buying, Selling, and Fundraising

ave you ever tried to get friends to take an old sofa off your hands? Or asked them to spread the word about a spare room you have for rent? Have you ever run a marathon for charity and requested donations? These types of social interactions have always happened on Facebook as well as out in the world, so after a while Facebook built some features to make these particular interactions easier and more streamlined. By using Facebook for these activities, you can reach a larger portion of your community instead of limiting yourself to just your friends.

Marketplace and Buy/Sell groups are your first stops for local buying and selling. Think of them as easy-to-browse yard sales managed by local community members or people interested in particular items. Marketplace covers virtually every type of thing you might want to sell, while Buy/Sell groups often have a specific focus. For example, you might join a Buy/Sell group centered on cookware, or one specifically for buying and selling snowboard gear. Depending on where you live, Marketplace might be more helpful than a Buy/Sell group or vice versa.

Fundraising is a way for you to solicit donations. It's a flexible system that allows you to fundraise on behalf of an established charity, yourself, or another person. News Feed is integral to getting the message out — countless examples exist of fundraisers that were widely shared and exceeded their initial monetary goals. For the purposes of this chapter, we assume that you're fundraising for an official non-profit organization.

Getting the Most Out of Marketplace

Marketplace is Facebook's central location for finding, buying, and selling new and used items. In recent years, Marketplace has expanded to include categories for larger things such as vehicles, rental properties, homes, and even job opportunities. It offers tools for searching and filtering your searches, as well as simple ways to keep track of items you've listed for sale.

Browsing and buying in Marketplace

To check out Marketplace, click the Marketplace link in the left sidebar on your Home page. This brings you to Marketplace, shown in Figure 12-1.

FIGURE 12-1:
See what's
in the market.

Marketplace has its own left sidebar and search functionality. The bulk of the page is taken up by listings. Each listing has a photo, a title, a price, and info about where it was posted.

The left sidebar has links to different sections: Browse All (where you start), Live Shopping, Notifications, Inbox, Cart, and Your Account. Below these are specific item categories. Clicking one of these categories often expands further subcategories. For example, when you click the Family category, you can fine-tune the results by clicking any of the smaller subcategories: Baby & Kids Items, Health & Beauty, Pet Supplies, and Toys & Games.

If you're looking for something in particular, use the search box at the top of the left sidebar to search for it by name or keyword. You can also change the city you're searching in under the Filters section. (Facebook autocompletes a city as you type the name of your desired location.) Use the drop-down menu to change the radius of your search. You can search as locally as within 1 mile of your city or town or as far away as 500 miles. By default, Facebook sets the radius at 40 miles.

After you choose a category, you can further filter by price range, delivery method (local pickup or shipping), and item condition (new or used). Other filters may pop up depending on what you're looking for. If you choose Furniture from the Home Goods category, you'll see a filter for Decor Style that includes options such as Art Deco, Farmhouse, or Bohemian. For Pet Supplies, you can choose a certain brand such as Purina or Kong. There's also a Free Stuff category for items that have been listed as free. This category can be a treasure trove. For example, if you're a gardener, the Free Stuff section is a great place to find plant cuttings. Try adjusting your filters if you're not finding what you're looking for.

Whenever you see an item that interests you, click it to open a larger image and view more details. You can see a sample listing in Figure 12-2.

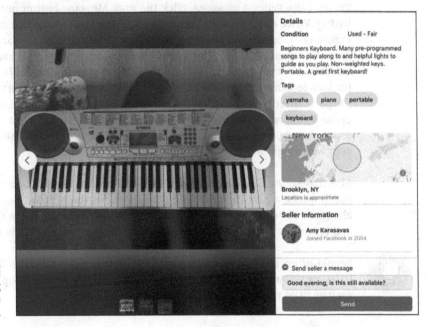

FIGURE 12-2: Checking out an item in Marketplace.

When you view an item's listing, you can click through the various photos of the item (if available). To the right of the photos you can view a more detailed description of the item, as well as a map indicating the general area it's being sold from. You can also view the seller's name and click through to view the person's public profile. This is one of those times where your sense of whether a profile is real comes into play. You don't want to start messaging back and forth with a fake or scammy account, so pay attention to when the seller joined Facebook (listed under the seller's name) and whether the person's profile seems fishy. Facebook suggests that you never include your email, phone number, or financial information in the first message to a seller, and that is sound advice.

WARNING

Some of the listings you come across may be ads in disguise. For example, nestled amongst the results of a search for Le Crueset cookware is an ad for a Le Crueset pan sold at a national home goods retailer. Luckily, there's an indication that this is an ad and not a regular listing. The word *Sponsored* appears where the item's location should be. Clicking the ad opens a separate tab or window for that item on the retailer's website.

You can take several actions from the item's listing. You can save the item to go back to it later. You can click to share the item with an interested friend. If you're ready to take the next step and inquire about the item, it's time to send the seller a message.

To send the seller a message, click the grey Message button near the top of the listing. Write any questions you have about the item in the text box, as shown in Figure 12-3, or use one of the prefilled text questions that appear as grey buttons above the message field. Click any of these commonly used phrases, such as "I'm interested in this item," to add it to the body of your message. We think it's a better idea to send a short and sweet custom message to help yours stand out from the pack. After typing your message, click the Send Message button to send it. This opens a chat window between you and the seller. If you don't hear back right away or you leave your computer, you can always find a record of the conversation in your Messenger inbox.

Alternately, you can use the big blue Send button at the bottom of the listing. You'll notice a prefilled text field directly above the button that says something along the lines of, "Hello, is this still available?" You can delete the existing text and type your own or send what's already written by clicking the Send button.

TIP

People often refer to messages between buyers and sellers as direct messages (*DMs*) or private messages (*PMs*). You might see this term pop up in descriptions such as "DM me for pricing" or "PM me for more info."

FIGURE 12-3:
Write to the
seller here.

Often the process of buying something requires a bit of a back-and-forth between you and the seller: You need to agree on a price and a method of payment, and coordinate a time and location for picking up the item or getting it delivered. The seller may be managing many incoming messages from others who are interested in the item as well, so be patient.

TIP

If you find sellers you particularly like, you can follow them. Head to one of their listings, click the seller's name, and click the Follow button. You will now be notified whenever that seller adds a new listing.

TIP

If you want to revisit a listing, save it. To go back to it, click Your Account and then Saved to see every listing you've saved. You can easily remove a saved listing by clicking the blue save icon. It will turn grey to signify it's no longer saved.

Live shopping

The live shopping model that QVC made famous is alive and well on Facebook. If you navigate to Live Shopping, you will see a list of videos that are live at that moment, with people selling everything from cosmetics to clothing to home goods. When you find the video you want, hover your cursor over it and click the play icon that appears. You can always switch to a different video or click the X in the upper-left corner to close the current video. Different sellers have different methods for purchasing goods over live video, which they will explain as you watch.

Selling your stuff on Marketplace

If you're trying to create a listing for something you want to sell, you can easily do so by going to Marketplace and following these steps:

1. **Click the blue + Create New Listing button in the left sidebar.**

2. **Choose a listing type.**

   Your options are Item for Sale, Vehicle for Sale, Home for Sale or Rent, and Job Opening. Job Openings operate a little differently than the other listing types; we explain how to post a job in the next section. After you choose a listing type, the Sell Something page appears, as shown in Figure 12-4.

FIGURE 12-4:
Create a listing
here.

3. **Add photos of your item by clicking the grey Add Photos button in the left sidebar and selecting them from your computer's hard drive.**

   Continue to click Add Photo if you have more than one. We recommend including at least a few photos because listings without photos receive less interest from buyers. Click the X icon to remove a photo.

4. **Enter the item's info into the appropriate spaces:**

   - *Title:* This is the headline for your item, so make it short and descriptive. For example, you might type IKEA Table and Chairs, Bundle of 3T clothing, Vintage iPad — whatever makes sense for the item you're selling.

   - *Price:* Decide how much you think your item is worth and add a price. If you aren't sure what price to enter, search for similar items already in Marketplace and see what other people are asking for them.

- *Category:* Choose an initial category from the drop-down menu and continue choosing subcategories (for example, clothing > women's clothing > women's dresses) until you run out of options.

- *Condition:* From the drop-down menu, choose New, Used — Like New, Used — Good, or Used — Fair. It can be challenging to be perfectly accurate here (one person's trash is another person's treasure and all), but try your best. If something is on its last legs, better to donate or properly dispose of it than to try to sell it on Marketplace.

- *Brand:* If the brand is an important part of the item's allure, be sure to add it here.

- *Size:* If you're selling clothing, shoes, or accessories, you'll see a text field to add the size of the item.

- *Description:* You don't need to fill out an additional description, but adding details about the item you're selling can be helpful. You can include its measurements, make any notes about use or care, or point out any wear and tear.

- *Product tags:* Tags are optional but they help refine searches and better match buyers with sellers. If you're selling a dress, add a tag for the brand, the size, the material type, and the fit. You can add up to 20 tags per item.

5. Specify the availability, if necessary.

On some listings, you'll be prompted to specify whether or not this is a single item or an item for which you have multiple versions. If the latter is the case, choose List as in Stock from the Availability drop-down menu. Most people are selling single items, but it's becoming increasingly popular for people to set up virtual storefronts on Marketplace.

6. Decide whether you want to hide this item from your friends.

You can determine whether or not your friends see that you've added an item to Marketplace. If you sell a lot of merchandise, you may not want to fill your friends' News Feeds with every listing. Turn on this setting to keep your Marketplace listings geared toward the community at large versus your personal network.

7. Click the blue Next button.

8. Add a location.

Facebook automatically populates your location, but you can change it by typing in the text box and choosing the correct city from the drop-down menu.

9. **Add delivery information.**

If you're an established seller, you may see an option to add delivery details. Decide if you'll limit your item to local pickup or if you'll also offer shipping. Click the Set Up Shipping button and follow the instructions to set up your account to receive payments.

10. **Click Publish.**

After a brief review, your listing is added to Marketplace, where everyone can see it and respond.

TIP

Click Save Draft at the top of the left sidebar to save your progress on a listing that isn't finished. You can return to an unfinished listing at any time by clicking Your Account and then Your Listings.

Keep track of your listings by going to the Your Account section of the left sidebar and then clicking Your Listings. Here, you can see every item you've listed for sale. When you sell something, click the blue Mark as Sold button to take it off Marketplace. Click the three dots icon to open a menu where you can edit the listing, mark the item as pending, or delete the listing. Facebook may also prompt you to boost your listing, which means paying to feature your listing as a Marketplace ad. It's up to you if you'd like to promote your listing in this way.

REMEMBER

All posts to Marketplace are public, which means everyone on Facebook can see them. Don't post any personally identifying information, such as your exact address, phone number, or credit card number.

TIP

For a comprehensive overview of your seller activity, head to your Commerce Profile from the left sidebar. You see a summary of your sales, including all active listings as well as your seller rating out of five stars (visible to the public after you've been rated five times). It's an easy way to see everything that you have for sale in Marketplace, in Buy/Sell groups, and on your timeline.

Posting jobs on Marketplace

In this section, we quickly go over the job functionality on Marketplace, which is different than buying and selling an item:

1. **Click the blue + Create New Listing button in the left sidebar.**

2. **Choose Job Opening as the listing type.**

3. **Post the listing as yourself or as one of your Pages.**

From the left sidebar, click the black down arrow to open a drop-down menu, and then choose to post the job as yourself or as one of your Page entities. We go over Pages in Chapter 14, but the general idea is that you can list the job as one of your businesses instead of listing it from your personal Facebook account.

4. **Enter the job's info:**

- *Job Title:* Make this title descriptive, such as "Social Media Marketing Manager for Large Fashion Brand."

- *Work Location:* Type the office address or the city where the job is located.

- *Job Type:* Click the field to open the drop-down menu and choose from Full-time, Part-time, Internship, Volunteer, or Contract.

- *Job Description:* This field is mandatory. Add some pertinent details about the duties of the position and any necessary skills. For example, does the applicant need to be proficient in a specific software program such as Microsoft Excel or QuickBooks?

- *Salary Range:* The minimum and maximum salary fields are optional as well as the field for whether the salary is hourly, daily, weekly, and so on. We recommend adding as much information as possible about the position to attract the best applicants.

- *Screening Questions:* Click the + Write Question button to add a screening question that will remove any candidates who aren't a good fit for the role. For example, you might ask something like "Are you willing to work on weekends?" When you click to add a question, you must first choose the question type from a drop-down menu with these options: Multiple Choice, Yes/No, and Free Text. Next, type your question in the question field. You can add as many of these questions as you want.

- *Receive Application by Email:* If you'd like to receive job applications by email, add your preferred email address here.

- *Photo:* Upload a photo to help your listing stand out. You might want to add a photo of the office or the last team-building event.

5. **Click Next.**

6. **Decide if you want to boost your job listing.**

Facebook gives you the option to boost your job listing, which will feature it more prominently on the jobs board. After you post the job in the next step, you will have to pay for this service and fill in more details.

7. **Click Post.**

Your listing is added to Facebook's public jobs board.

8. **Manage your listing.**

   You can access your listing any time by navigating to Jobs from the left sidebar of the Home page and clicking the Manage Jobs button. Note that this is where you will also find submitted applications for listed positions. Click a listing to take actions such as closing the job, sharing the job, viewing the listing as an applicant, and deleting the job.

Marketplace inbox

Facebook knows you don't want to crowd your inbox with buyer inquiries, so they created a unique messenger thread for inquiries about your listing. The thread, labeled Marketplace, exists in your inbox. Clicking it opens a list of individual messages about the items you are buying or selling. This way, all Marketplace messages are in one central location and you don't need to keep track of them amongst your personal messages. Note that some accounts can access these messages by clicking Inbox from the Marketplace's left sidebar. This feature is not yet available to everyone at the time of this writing, so don't worry if you don't see it yet. You can still access your Marketplace messages by heading to your regular Facebook Inbox.

Using Marketplace on your phone

 Marketplace is easy to browse and use on your phone. To get to it (assuming you're using the Facebook app on an iPhone or Android), tap the Marketplace icon at the bottom of the Home page. Scroll up and down to browse. Tap any listing you find interesting to learn more.

You can quickly tap the blue Send button to send the prefilled message "Hi *<seller name>*, is this still available?" or compose a message of your own. You may also click the grey Send Offer button to offer the seller a particular price. Keep in mind that offers are not payments and that purchase details will still need to be arranged with the seller. You may also see buttons like Save, Share, or Buy Now from established and highly rated sellers.

Using your phone to create a listing is convenient because you can easily take a picture of the item you want to sell without having to search your hard drive or transfer photos from your phone to your computer:

1. **Tap the grey Sell button at the top of the page.**

   A menu of categories for your listing appears.

2. **Choose from Items, Vehicles, Homes for Sale or Rent, or Jobs.**

All categories except Jobs open the Sell Something page.

3. **Tap Add Photos to open your phone's camera roll, and select your photos.**

You can also click the camera icon to open your phone's camera and take photos.

4. **Enter the item information (title, price, category, description, tags, and so on) in their respective fields.**

Some additional options here are not on your computer. You can be more precise with your description, for example, adding style, length, and occasion to a listing for a dress.

5. **Note the availability, if necessary.**

On some listings, you'll be prompted to specify whether or not this is a single item or an item for you have multiple versions. If the latter is the case, choose List as in Stock from the Availability drop-down menu.

6. **Designate a door drop-off or a pick up, if necessary.**

Due to Covid-19, Facebook has added a check box to inform buyers when you prefer a drop-off or pick-up at the door. This option may be located in different places on the listing depending on your account, so don't worry if it doesn't appear on the first screen.

7. **Decide whether you want to hide this item from your friends, and then click Next.**

You can determine whether or not your friends see that you've added an item to Marketplace. If you sell a lot of merchandise, you may not want to fill your friends' News Feeds with each and every listing. Turn on this setting to keep your Marketplace listings geared toward the community at large versus your personal network.

8. **Add location information and delivery method and then click Next.**

This option appears only for established sellers. Most people choose to limit their listings to a local pickup to save a trip to the post office and remove shipping costs as a factor. If you choose Local Pickup Only, you may see an option to limit the exchange of goods to the doorway in compliance with Covid-19 best practices.

9. **If you want to share this item in any of the Buy/Sell groups you belong to, select the box next to each group's name.**

10. **Tap Publish.**

Your listing is added to Marketplace.

BUYING AND SELLING SAFELY

While Facebook tries to take some of the guesswork out of buying and selling used items online, the fact remains that you'll be interacting with someone you don't know. Here are a few basic safety tips to keep in mind:

- **Watch out for scammers.** If something sounds too good to be true, it probably is. Don't let people upend your common sense. There is no reason someone selling something or buying something from you will ever need your password, credit card number, or anything like that. You should never transfer money to someone until you have the item in your hands and have checked to make sure it is as advertised.

- **Do your due diligence.** Using Marketplace to find things like the perfect vase or a gently used toy is all well and good, but people also sell much larger items, such as vehicles and homes. Big-ticket items require more due diligence on the part of the buyer. Make sure the seller is reputable and follow best practices such as asking for and checking references, looking at the person's seller rating, and comparing similar listings.

- **Use cash or person-to-person payment methods.** You can make payments using Facebook or another app, in which money is moved from one account to another in a few minutes, such as Venmo, Zelle, or PayPal's Instant Transfer service. Checks can be faked or bounced and money orders aren't always reliable. Cash is probably your best bet.

- **Try to meet in a neutral location.** If possible when arranging to sell someone something, try to meet in a neutral, public location. This makes both people feel safer about meeting someone from the Internet. Let someone know where you'll be and who it is you are meeting.

- **Give out your address judiciously.** Of course, a neutral location might not be possible or worthwhile, especially if you're selling something that's large or heavy. So don't give out your full address to someone until they've committed to coming to pick up your item. And if something about the person makes you uneasy, don't force yourself to meet him or her!

Don't let the number of steps here intimidate you. We once listened to someone describe the process of listing multiple items to Facebook as something that was easily accomplished with a tasty beverage in one hand and a phone in the other.

Belonging to Buy/Sell Groups

In addition to Marketplace, many people buy and sell items in designated Buy/Sell Groups. You can use both (and, in fact, Facebook makes it easy to post Marketplace listings to Buy/Sell groups you belong to and vice versa). People tend to use Buy/Sell groups for the following reasons:

>> **Groups require membership:** Members of a group must actively join, and this requirement acts as a filter on both the number of people who will see your listing and the type of buyer/seller you are interacting with. Not to say that the Buy and Sell safety tips don't apply — just that you're sharing listings with a particularly interested group of people.

>> **Groups can get specific:** You may only want to search for items available within a small radius, perhaps 5 miles or less, from your home. Buy/Sell groups often limit themselves to certain neighborhoods. They can also be for specific categories of items, such as outdoor gear, collectibles, and housing. This type of targeted browsing can be a time-saver and a sanity-saver.

>> **Groups often have a trade element:** In addition to selling more locally and in a community-based way, many Buy/Sell groups offer an option to trade items. If you're short on cash or hoping to get rid of a few of your own possessions, the barter system might be what you're looking for. Beyond this, keep an eye out for local "Buy Nothing" groups, built around the minimalist notion that we don't need to buy anything new, ever. In those groups, members can create requests for items, and other members, fulfill those requests if they can.

The mechanics of using Buy/Sell groups are similar to any other group. You use the share box, comment on posts you're interested in, and message other people in the group. If you need a refresher on how groups work, check out Chapter 10.

To find Buy/Sell groups near you, click Groups in the left sidebar on the Home page. Then click Discover, and under Categories you'll find Buy & Sell. Clicking this shows you the popular groups in your area. Click the Join Group button for any group that looks appealing.

Browsing and buying in a Buy/Sell group

As shown in Figure 12-5, a Buy/Sell group looks pretty much like most other groups. The main difference is that the share box defaults to Sell Something instead of prompting you to start a discussion. Under the share box are usually a few announcements from the admins as well as top listings. Click the down arrow next to Top Listings to change the view to New Listing Activity or Nearby Listings. These listings should look familiar to you — they're just posts, albeit with titles and prices. Some posts will have comment threads where people ask questions about the item's condition, pick up location, or other related topics.

FIGURE 12-5:
Buy/Sell Groups.

In general, people don't use comment threads to perform the final negotiations for buying and picking up an item. Instead, when they're ready to make an offer, most people message the seller directly (using a handy Message button in the lower-right corner of the post).

If you're on the hunt for something, click the search icon (magnifying glass) near the top right of the page to search for that item in the group's posts.

Selling items in a Buy/Sell group

If you want to list something for sale in a Buy/Sell group, click the share box. A little text box under the cover photo contains the Sell Something button. When you click it, the Item for Sale dialog opens, as shown in Figure 12-6.

Does this screen look familiar? It should because it's a more concise version of the screen shown in Figure 12-4. You'll need to fill out the following fields:

» **Photos:** Add photos of the item you're selling. Generally, it's a good idea to add more than one photo.

» **Title:** Your answer to this question is the headline for your post, so be descriptive.

» **Price:** Let people know how much they need to pay for the item.

» **Description:** Add any details about your item that may be relevant, such as condition and measurements.

» **Product tags:** Create up to 20 tags to make it easier for people to find your item.

» **Location:** Facebook may automatically fill in your city. Delete this info if you need to enter a different city.

FIGURE 12-6:
Create your listing here.

Click Next to advance to the next screen and choose whether or not you'll add this listing to Marketplace or other Buy/Sell groups that you belong to or both. When everything is to your satisfaction, click the big blue Post button. Your listing is now live in the group and wherever else you decided to post it. You'll be notified about any comments on the listing, as well as any incoming messages from people who are interested in buying your item. You can track all your listings in a single group by heading to that group's Your Items tab.

Using Buy/Sell groups on your phone

Much like using Marketplace, using Buy/Sell groups on your phone — particularly for selling items — can be even easier because you can take pictures directly from your phone and include them in your listing. Take it from us, if you're sitting at your computer trying to create a listing, you will inevitably need to make many trips back to the item to check the measurements, take a new photo, and so on. Make your life easier and create the listing from where the item is located.

To navigate to the Buy/Sell group you want to use for your sale, type its name in the search bar at the top of the app. You can also tap the three dots icon in the bottom-right corner to view a menu of all the shortcuts, features, and destinations you can go to in Facebook. You can click your desired Buy/Sell group there.

Fundraising for Causes

Facebook has always been a place where people can drum up support for a cause. People who participate in a fundraising effort such as Breast Cancer Awareness often reach out to their Facebook network for donations and support. In this way, fundraising on Facebook is nothing new, but the tools now available make it much easier to get support.

Donating to a fundraiser

You might learn about fundraisers from an invite or see that one of your friends has created or donated to a fundraiser in your News Feed. A sample post is shown in Figure 12-7.

FIGURE 12-7:
Your friend is
raising money!

Much like any other post you might see in your News Feed, these posts have options to like, comment, and share. In addition, they also have a Donate button in the lower-right corner of the post, which you can click to initiate a donation.

If you want more info about the fundraiser, click the fundraiser's cover photo or title to go to the fundraiser's page. There you can read the full story about the fundraiser — why your friend has started it, what exactly your money will be accomplishing, and so on. If someone is raising money for a nonprofit, you might want to double-check to see if what your friend says the money is for is the same thing the nonprofit says the money is for.

Clicking Donate on either your friend's post or the fundraiser's page opens the donation window shown in Figure 12-8. In the top part of this window, choose the amount you want to donate and decide if you want to donate this one time or set up a monthly donation. Enter your credit card information or log in to your PayPal account to set up the payment.

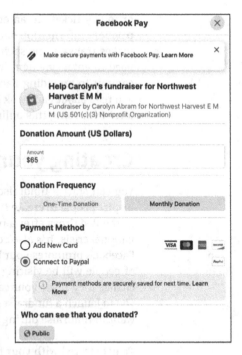

FIGURE 12-8:
Donate here.

By default, donations are public, though the amount of any donation is kept between you and the person organizing the fundraiser (and the eventual recipient of the donation). You can also choose to keep your donation visible to only friends or only yourself, but the organizer and recipient will always be able to see that you have donated and the amount. When you've entered all the necessary info, click the Donate button.

REMEMBER

Sometimes you'll feel compelled to donate to a fundraiser and other times you might just want to offer support by leaving a public comment on the News Feed post or sharing the fundraiser to your own network to amplify its reach. There is no one right way to respond to a fundraiser.

Facebook Pay

Facebook Pay is Facebook's product for moving money between users. You can use it to pay for goods and services on every Facebook-owned platform, including Instagram and WhatsApp. You can use Facebook Pay in a lot of ways on Facebook itself: to send money to a friend in Messenger, to purchase items on Marketplace, or to buy tickets to an entertainment event. It's easy to sign up. You enter your payment information once, and Facebook remembers it for future purchases. This makes transactions such as donating to fundraisers seamless because you don't need to enter payment information every time you donate. However, before you commit to providing Facebook with your payment information, we suggest that you read about Facebook Pay at http://pay.facebook.com to gauge your comfort level with using this online payment application.

Creating your own fundraiser

You can start a fundraiser for any reason at any time. Often current events will inspire people to create fundraisers — whether that's linking to a big charity such as the Red Cross after a natural disaster or donating to a local food bank during a national crisis. It's also common to create a fundraiser for one's birthday. Because Facebook promotes your birthday to your friends, it's a time when you know a lot of people will be visiting your timeline and leaving a message, so there's a good chance they'll see your call to give as well. In fact, Facebook may actively prompt you to launch a fundraiser close to your birthday. It's up to you whether or not you want to fundraise during this time.

To get started with your fundraiser, follow these steps:

1. **On your Home page, click Fundraisers in the left sidebar.**

   You see the Fundraisers page, which includes more information about fundraisers and a list of any fundraisers your friends are currently running.

   You must be at least 18 years old to create a personal fundraiser.

WARNING

2. **Click the + Raise Money button in the left sidebar.**

3. **Choose the type of fundraiser.**

   You can raise money for a nonprofit, a personal cause such as a friend or business, or yourself. If you choose a nonprofit, you select the nonprofit company from a list that Facebook provides. If you choose a personal cause or yourself, you add context by choosing a category such as Medical, Personal Emergency, or Business.

4. **Fill out the information on the Raise Money page.**

   The Raise Money page, shown in Figure 12-9, looks a lot like the page for creating an event or creating a group.

5. **Double-check (or create) a title for your fundraiser.**

   Titles should be brief and descriptive, such as "John Smith Memorial Fund," "Help Tyler Pay for Knee Surgery," or "Amy's Fundraiser for RAINN."

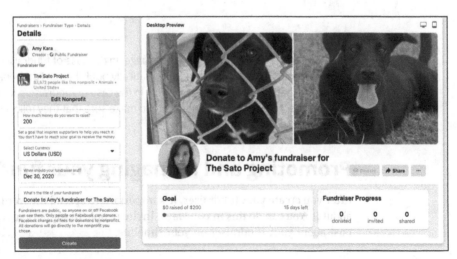

FIGURE 12-9:
Create your
fundraiser here.

6. **Tell your story.**

   Use as much space as you need to explain why you're creating the fundraiser and what the money will go towards. This is a great place to anticipate any questions your friends may have about the fundraiser: Explain why you are fundraising, why you were moved to donate, how the person or organization will be receiving the money, and any other information you think might be relevant.

7. **If you're fundraising for a personal cause or yourself, set your country.**

   This should be the country where your bank is located so that the funds can get to you properly. Choosing a country will lock in the currency type as well.

8. **Set a fundraising target.**

   It's usually better to set a lower target at first and continue increasing it whenever the goal is surpassed.

9. **Pick an end date for your fundraiser.**

   Having an end date helps your friends know how long they have to donate and lets you issue reminders such as "only three days left!"

10. **Choose a cover photo.**

    Facebook recommends default cover photos for your fundraiser depending on the organization or the category. You can use what Facebook recommends or click the adjust photo icon (two photos and a plus sign) in the upper-right corner of the cover photo preview to choose a different option or to choose a photo from your computer.

CHAPTER 12 **Buying, Selling, and Fundraising** 303

11. **Click Create.**

This brings you to your fundraiser's page. In the case of fundraisers for nonprofits, your fundraiser is automatically published to Facebook. (Keep in mind that all fundraisers are public.) In the case of fundraisers for individuals, Facebook reviews such fundraisers to make sure that they follow Facebook's standards.

Promoting and managing your fundraiser

After you create your fundraiser, it's important to spread the word so people know about it. You can do this in two ways: Invite people to support your fundraiser, and post about your fundraiser on your timeline.

Inviting friends to support your fundraiser

To invite friends to view and support your fundraiser, follow these steps:

1. **From your fundraiser's page, click the blue Invite button (below the cover photo and title).**

This opens a window with a list of your friends.

2. **Click the blue Invite button next to any friend's name to invite him or her to view your fundraiser and (you hope) donate.**

You can use the search box at the top of the window to search for specific friends.

3. **Click the X in the top-right corner of the window to return to your fundraiser page.**

Friends you've invited receive a notification about the fundraiser.

Sometimes people wonder which friends they should directly invite to a fundraiser. On the one hand, if you're fundraising for a cause you believe in, you should reach out far and wide; on the other hand, people can feel that it's rude to ask their friends for money constantly. We don't have a perfect answer for this question, other than to recommend that you invite the same people you would reach out to via direct email or phone call for a fundraising effort. We'd leave out any coworkers (especially if you're their boss!) or people who might feel awkward declining to donate.

Creating posts about your fundraiser

Facebook will prompt you to invite friends and post about your fundraiser after you've launched it. They also suggest donating to your own fundraiser first because people are more likely to donate if they see that the ball is rolling. If you choose not to post a story immediately after creating your fundraiser, you can do so at any time by clicking the Share button on the fundraiser's page. This opens a window for creating a post that is the same as the share box, but the information about your fundraiser is already filled in. In the text box above the fundraiser's cover photo, add any text you want to include about why you started the fundraiser or why you want people to donate, and then click Post.

Remember, fundraisers are public, so even if you only share your post with friends, those friends will be able to share the fundraiser with their own friends if they are so inclined.

Getting the money where it needs to be

If you're running a fundraiser for an established nonprofit, Facebook automatically links to their online payment systems to route donations to that organization. However, if you're creating a fundraiser for yourself or a friend, you need to make sure that the money goes to the right place. Facebook uses Stripe, a credit-card processing company, to run these transactions. You need to know the account number and routing info for the account where you would like the money deposited. You may also need to submit further information to prove your identity.

Unfollowing, editing, ending, and deleting a fundraiser

From the fundraiser's page, clicking the three dots icon (located below the cover photo and title) opens a menu with several options:

>> **Unfollow:** After you donate to a fundraiser, you may see updates from that fundraiser in your News Feed. If you'd rather not see those updates, you can unfollow the fundraiser by selecting this option.

>> **Edit:** As fundraisers you've created go on, you might find that you want to edit some detail. For example, you may decide to add an FAQ section to the story or increase the amount of money you want to raise. Selecting Edit from the three dots icon opens the Edit Fundraiser window, where you can change most of the fields you filled out when you were creating the fundraiser. Remember to click the Save button if you make any changes.

>> **Match Donations:** If you find that your fundraiser isn't performing as expected, you have some ways to up the ante. One is to pledge a donation match. You set an amount that you feel comfortable matching to encourage your friends to give. For example, you might say "If all my friends donate $1,000 to breast cancer research, I will match it with my own $1,000."

>> **Add Organizers:** Perhaps you're running a large fundraising effort and you could use some help. From here, choose up to three Facebook friends to assist you with editing the fundraiser and managing posts. They will be notified of your request that they become an organizer and can choose to accept the role or decline.

>> **End:** If you choose to end your fundraiser, for whatever reason, the fundraiser will still exist as a page on Facebook, and you can still post updates and view any info there. This feature is significant for certain fundraisers, such as a memorial fundraiser, where the families might take comfort in being able to go back and view the messages people have left for them after a loved one's death.

>> **Delete:** When you delete a fundraiser, all of its info, including any messages people have left, is deleted from Facebook. Nonprofits that are associated with the fundraiser will still receive any donations that have been made so far. Note that you must first end a fundraiser to delete it.

IN THIS CHAPTER

» **Checking out public events**

» **Hosting an in-person versus online event**

» **Responding to event invitations**

» **Creating and managing your own events**

Chapter **13**

Scheduling Your Life with Events

acebook events work well for the same reason lots of other Facebook features
work: Your friends are here on Facebook. Event pages act as a hub for gatherings of any kind, whether online or in-person. You can invite friends to
events, keep track of RSVPs, and use Facebook to send updates or coordinate
participation — how else will everyone know which dish to bring to the potluck?
You'll also use Facebook's Event Calendar to keep track of upcoming events and
your friends' birthdays.

You're Invited!

You'll most likely find out about an event through a notification. When a friend
invites you to an event, a small red circle appears over the notifications icon (the
bell icon in the top bar). Click the icon to open your notifications menu; then click
the invitation to be taken to the event. You can go to your Events page by clicking
Events in the left sidebar.

A sample event appears in Figure 13-1. The event photo is at the top of the page,
much like the cover photo on your timeline. Below the event photo are the event
details, including name, location, time, host (which can be a person, group, or
Page), and RSVP options.

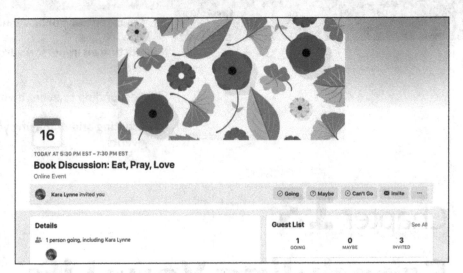

FIGURE 13-1:
An event on
Facebook.

Events have four possible privacy settings:

>> **Public:** Public events are just what they sound like — open to the public. Anyone on or off Facebook can see the event, view the guest list and posts, and join the event. No invite needed!

>> **Private:** Only people who have been invited are able to see the event and join it. Depending on the event creator's settings, invitees may be able to add more friends to the event as well. This setting also allows the creator to invite people who aren't on Facebook.

>> **Friends:** Similar to private events, but the creator is able to invite only their friends on Facebook.

>> **Group:** The creator can specify an event for a specific group they belong to. For example, a book club group may create an event to discuss the latest book they're reading.

To see the most important info about an event, head to the Details box in the center of the page. There's a spot here for the date and time of the event, the location, and any info the event creator wants to share with guests. If it's an in-person event and the creator listed a specific address or location, you'll see a map and possibly a weather forecast to the right of the Details box. You can enlarge the map by clicking it.

To RSVP to an event, look below the event photo where you can see who invited you to the event alongside three buttons: Going, Maybe, and Can't Go. Click the appropriate button so your host can get an accurate head count. You can see other people's RSVPs in the Guest List section.

Depending on the event, you may see an Add a Post button below the Details box or you may see two tabs: an About tab with event info and a Discussion tab with the Add a Post button. These are designated spaces where people can post messages and communicate with other guests (and potential guests), as shown in Figure 13-2. Event hosts use this section to post important updates — for example, a reminder to bring a sweater in case the weather turns cold, or info on where to meet in the park. Posts may be from guests expressing excitement, coordinating rides, or explaining why they'll be showing up late.

FIGURE 13-2:
Post here to communicate with other guests.

TIP

Certain kinds of events — those created by a group or a Page — may require posts to be reviewed by an admin before they appear to other guests. Hosts may also opt to let only admins post, which removes the share box from the Discussion tab. This is more common for large public events.

To post to an event, follow these steps:

1. **Click Add a Post to display the share box.**

   This is the same way you update your status or post to a group, except in this case your post will be shared with people who can view the event.

2. **Type your message in the share box.**

3. **(Optional) Add photos, tags, activity, or location information by clicking their respective icons at the bottom of the share box.**

4. **Click Post.**

   Your message is posted to either the home page or Discussion tab of the event. Depending on settings, guests may be notified about it or see a News Feed story.

One feature of events is the ability to create a poll that all members of the event can respond to. This can make some of the planning that goes into certain kinds of events a bit easier. Want to know how many people will eat pizza if you order it? If anyone has any food allergies you should know about? You can create a poll to ask them. Just follow these steps:

1. **Click the three dots icon in the share box.**

2. **Click Poll.**

3. **In the Write Something box, type your question.**

4. **In the Add Poll sections, enter possible answers to the question.**

   Although three option fields are visible, you can add almost as many options as you can think of. To add option sections, click the +Add Option button. As you keep clicking to add more options, more fields will appear.

5. **While creating your poll, click the Poll Options button at the bottom of the share box and decide on the following two options:**

   - *Allow people to choose multiple answers:* By default, polls allow people to choose as many options as they like. If you'd rather people choose only one option, deselect this box.

   - *Allow anyone to add options:* By default, if poll participants don't see an answer that applies to them, they can create an answer option. You can turn off this option if you'd rather they be forced to choose one of the options you provided.

6. **After you add all the options you want, click Post.**

   Everyone who is invited to the event can then respond. Keep in mind that some events are public, in which case an unlimited number of people can respond.

Public Events

Most of the time, you'll view an event because you were invited. But occasionally, especially for large-scale public events, you may see an event in your News Feed. These events look the same and have the same components as smaller, private events, but the options for RSVPing are Interested and Going (instead of Going, Maybe, and Not Going).

When you're viewing the event, you can click the Going button to RSVP to the event, or click the Interested button to see reminders about the event and receive updates from the event. In other words, Interested is the equivalent of subscribing or following someone. When you list that you're interested in or going to a public event, that may create a News Feed story that your friends could see.

TIP

After you RSVP, whether you replied Going or Not Going, the RSVP buttons disappear and you see one button that shows your attendee status. If you need to change your RSVP, click that button to open a drop-down menu of every RSVP option.

Viewing Events

From your Home page, you can see reminders about upcoming events in the Shortcuts section of the left sidebar. (Friends' birthday reminders are on the right side of the page.) Click Events in the left sidebar to visit event central. The left side now contains options related to events.

>> **Home:** Get an overview of events happening now, options for browsing events, daily recommendations, and suggested events based on your interests.

>> **Your Events:** An overview of your upcoming events listed by date. These are the events you've been invited to, RSVPed yes to, and said you are interested in. From the left sidebar, click the down arrow to the right of Your Events to see lists of pending invitations, upcoming events, events that you're hosting, and past events. Figure 13-3 shows some of Amy's past events.

>> **Birthdays:** See when your friends' birthdays are (assuming they share this info).

FIGURE 13-3:
Reflect on your busy social life.

>> **Categories:** Check out events in your area or online that you may be interested in by choosing from a plethora of categories, including Art, Fitness, Religion, and Networking. You can filter events by time, date, location, and whether or not the event is online.

TIP

If you're the kind of person who likes to visualize your schedule on an online calendar, such as Google Calendar or Microsoft Outlook, click the three dots icon in the top-right corner of the event and choose Export Event from the drop-down menu. (If you're using Facebook on your phone, you'll need to open the event to access the three dots icon and choose Add to Calendar from the menu.) Choose Save to Calendar or Send to Email, and then click the blue Export button to download the event to your computer or phone, making it easy to add to your calendar of choice.

Creating an Event

Eventually, the time may come when you want to organize an event. It might be a party or a barbecue or a book club or any other gathering of your friends. No matter what the context, follow these steps to create your own event:

1. **Click Events in the left sidebar of your Home page.**

2. **Click the blue +Create New Event button at the bottom of the left sidebar and decide whether the event will be online or in-person.**

   The Event Details page appears, as shown in Figure 13-4.

FIGURE 13-4:
Create your event here.

3. **Fill out your event's info.**

You can fill out a number of fields. Click the blue Next button to advance to the next screen:

- *Event Name:* Fill in the name of the event, not your name (a common mistake). Usually events get descriptive names such as Amy's 30th Birthday Bash or Brooklyn Heights 5K.

- Start *Date and Time:* By default, Facebook assumes you're an impromptu party planner, so the date in the box is today. Click the calendar icon to change the date. Next to the date box is a box for the event's time. Type the time your event begins. Click +End Date and Time to let folks know when the fun ends.

- *Privacy:* Choose one of the four options we mention earlier (Public, Private, Friends, or Group). If you choose Private, you'll be able to turn off the ability for guests to invite their friends and risk strangers crashing your intimate dinner party.

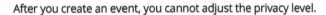

  After you create an event, you cannot adjust the privacy level.

- *Location:* Events are better when people know where to go! You can type an address or a location (such as a restaurant or a park). Facebook attempts to autocomplete a specific location while you type. When you see the desired location, click it or press Enter. If you're hosting an online event, choose a virtual location, including a Facebook Live video (see Chapter 11 for more), an external link, or a destination of your choosing. Just be sure to make instructions explicit and easy to follow.

- *Description:* (Optional) This is the info guests will read about when they see the event, so provide any info that helps people understand what they're going to. For example, a bookstore event might list the readers who will be in attendance, or a 5K run will include the location of the starting line.

- *Cover Photo:* Events look nice with a pretty picture to go along with them. Click Upload Cover Photo to choose a photo from your computer's hard drive. Alternately, you can choose Illustration to select from one of Facebook's many prefab themes.

- *Event Settings:* Click the gear icon to add co-hosts by typing the name of any of your Facebook friends. (The invite to serve as co-host won't be sent until the event is created.) You can also turn off the setting that displays the guest list to other guests. More often than not, event creators leave this setting on so that guests have an idea of who they'll be mingling with.

4. **Click the blue Create Event button.**

You're taken to your event's page. Even though your event is created, you still need to make a few finishing touches, such as actually inviting some people.

REMEMBER

Some settings in Facebook's mobile app are different than the desktop version. For example, while creating an online event on a mobile device, you have the additional option to hold your event in Messenger rooms. (See Chapter 9 for more about Messenger rooms.)

Inviting guests

Trust us, your party just won't be the same unless some people show up, and the number-one way to get people to show up is to invite them. Inviting people to your event doesn't have to be a one-shot deal. You can follow these steps at any time to invite people to your event:

1. **From your event's page, click the Invite button in the upper- right portion of the page (under the event photo).**

    The Invite window appears, as shown in Figure 13-5. If you have a private or public event, you can search for and choose to invite Facebook friends, or invite friends who don't use Facebook. Friends and Group events are limited to Facebook users.

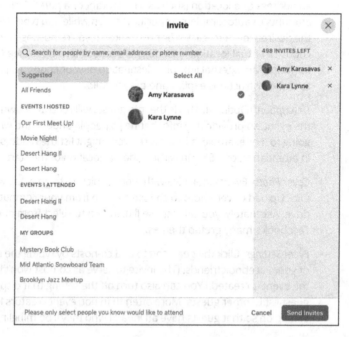

FIGURE 13-5:
Be our guest!

2. **If you chose to invite Facebook friends, click a friend's name or face to select him or her.**

Because you may have a lot of friends, you can use the search box at the top to search for people by name. You can also use the categories on the left side of the window to filter to certain friends. For example, you may be able to look at friends who live near you, or friends who are in a group with you. After you've selected a friend, he or she appear on the right side of the window under the number of invites that you have left to send. (Facebook caps event invites at 500 per person.) If you're looking at a list of your friends and realize you want to invite everyone on the list, save yourself some time and click the blue Select All link to add everyone at once.

TIP

Each co-host can give out 500 invites, so you can get around the 500 invites cap by naming a Facebook friend as the event's co-host.

3. **If you chose to invite via text or email, type a mobile number or email address, respectively, in the search bar at the top of the window.**

Facebook will try to match you to a Facebook profile linked to that mobile number or email address. If they match, tap your friend's thumbnail photo to select the person. If there's no match or you're not sure, you can just choose to send the invite directly to that email or phone.

4. **Click the Send Invites button after you make all your selections.**

At this point, your guests receive the notification of a new invitation and will probably begin to RSVP.

Managing Your Event

After you set up your event and the RSVPs start trickling in, you may need to manage some things. You might need to provide more info or change the location to accommodate more people. If it's a large online event, you may need to moderate the people posting. Here are some common management issues you might face and how to deal with them.

Editing your event's info

Need to update the start time or add info about a dress code? You can do so at any time by clicking the Edit button below the Event's cover photo on the right side of the page. This opens the Edit Event Info box, which looks exactly like the Create New Event box. You can change the name of the event, the date and the time; add more event details; and change the location. You can also add more hosts to the event (by default, as the creator of an event, you're already its host). Hosts have the same capabilities you have in terms of editing the event.

Click Save when you're finished editing.

If the event is a repeating one, such as a weekly bingo night, you can make life easier by clicking the three dots icon to the right of the Edit button and choosing Duplicate Event from the drop-down menu. You can then repost the event with a different date, time, or location, without changing the other settings.

Canceling the event

If plans have gone awry, not to worry — it's easy to cancel your event:

1. **From your event's page, click the three dots icon and then choose Cancel Event.**

A window appears asking whether you're sure you want to cancel the event.

2. **Click one of the following:**

- *Cancel Event:* Canceling an event tells people that the event has been canceled but leaves the event as a page on Facebook where guests can communicate with each other.

- *Delete Event:* Deleting an event tells people that it has been canceled and removes from Facebook anything that's been posted to the event. Choosing this option makes it crystal-clear to guests that the event is off.

3. **(Optional) In the Say Something box, write a quick explanation of why you're canceling.**

4. **Click Confirm.**

The event is immediately canceled, and notifications will be sent to guests letting them know.

Messaging your event's guests

The most common way of communicating with your guests is simply to post something to the Discussion or Posts section of the event. Click Add a Post to open the share box, type your message, and click Post.

In addition, you can message guests individually:

1. **Navigate to the Guest List section of your event.**

2. **Click the Message button to the right of the guest's name, as shown in Figure 13-6.**

   A Messenger chat window appears.

3. **Type your message and press Enter or click the blue arrow icon.**

   Your message becomes part of your Messenger history with that guest and he or she can reply at will.

FIGURE 13-6:
Messaging guests.

Removing guests

Although removing guests isn't something that happens often, if you're hosting a large event (say, a big public fundraising effort for a charity), you may find that certain guests are undesirable, especially in the Discussion section of the event. You can remove inappropriate posts (as well as reporting spam or abuse should that happen). If there's one bad egg, you can remove that person from the event as follows:

1. **In the Guest List box, click See All.**

   The Guest box opens.

2. **Click the three dots icon next to the name of the person you'd like to remove from the event.**

   The person is removed from the event but will not be notified about the removal. This is not a permanent ban; you can resend an invite if you remove a person accidentally.

Chapter **14**

Creating a Page for Promotion

Picture your town or city. There's the park, the school, the houses where people live, the offices where they go to work, and the stores they shop in. When you drive around town, you see all sorts of activities happening — whether people are walking their dogs, grabbing a cup of coffee, or working out. The world we live in is comprised of people, the stuff they do, and the stuff they need or want. People have real connections to all this stuff: the shops, the brands, the bands, the stars, the activities, the passions, and the restaurants and bars — things that are important. Facebook is about people and their real-world connections, including the connections that aren't friends. On Facebook, these non-friend entities are represented as *Pages*.

REMEMBER

Page is a common word on the Internet, so we always capitalize the *P* in Pages when talking about Facebook Pages.

This chapter is all about understanding the world of Facebook Pages. If you just want to understand these things that you've been liking and that have been showing up in your News Feed, check out the next section, "Getting to Know Pages." If you're looking to represent your business, brand, band, or anything else on Facebook, continue to the "Creating a Facebook Page" section.

Getting to Know Pages

You can do many of the same things with Pages that you can do with friends — write on their timelines, tag them in status updates, and receive their content in your News Feed. The main difference is that instead of friending Pages, you like them.

When you like a Page, your information and access to your timeline isn't shared with the Page and its admins, although any posts you make to a Page are public.

REMEMBER

Facebook recently announced upcoming changes to the Page format to focus on follows instead of likes. This change will occur over time, so don't worry if your Page still has a Like button; at some point it will change to a Follow button, which will make it easier for your fans to engage with your Page without having the extra step of deciding whether to like or follow. As of this writing, however, the instructions regarding how to like or follow a Page still apply.

Anatomy of a Page

Pages are meant to resemble profiles and timelines, so if you've already read Chapter 4, most of the following will sound familiar. Figure 14-1 shows a sample Facebook Page from *For Dummies*.

FIGURE 14-1:
The *For Dummies*
Facebook Page.

Here's the anatomy of a Page, across the top from left to right:

» **Profile picture and cover photo:** Just like you and your friends, Pages use one photo to represent them across the site. Usually, it's a logo or an official press photo. Pages also have a cover photo, often an image that more broadly represents the organization's brand. A blue check mark to the right of a Page name indicates that it's a *verified Page*, meaning that Facebook has confirmed that this Page is indeed managed by the company it represents.

» **Call-to-action button:** To the right of the Page name is usually a big blue button with an action you can take. A common button is Send Message, but you may also see Sign Up, Shop Now, or Learn More.

» **Page tabs:** Under the profile picture are different tabs that you can navigate to. By default, you land on the Home tab. Head to another tab to see more Page content, such as the Photos tab, which contains every photo the Page has ever uploaded, or the Events tab, which highlights any events connected to the Page. The Community tab has information about how many people have liked the Page, including a list of any of your friends who already like the Page. You may need to click More to see additional tabs.

» **Like button:** The Like button is located to the right of the Page tabs. To connect to a Page, click this button. (See the "Connecting and interacting with Pages" section for more on this topic.) If you've already liked a Page and want to unlike it, click the blue Liked button, which changes it to a grey Like button.

» **Message button:** You may see a Message button next to the Like button. Click this to send a message to the Page admin(s).

We refer to people who are connected with Pages as *fans* of the Page.

» **Search icon:** Click the magnifying glass to search for particular posts or keywords. Perhaps you want to know if the local coffee shop also sells donuts. Go ahead and type donuts into the search bar and click Enter on your keyboard to find out.

» **More icon:** Clicking the three dots icon opens a drop-down menu from which you can take many actions. The first is to adjust your follow settings. When you like a Page, you automatically follow it and receive its posts in your News Feed. Your follow settings are similar to your News Feed settings for friends. For example, you can snooze the Page in your News Feed for 30 days or you can turn off following but still like the Page. Inversely, you may want to follow a Page without liking it when you're interested in the content the Page produces but don't want to be associated with liking that content. Like implies approval and not everyone is comfortable giving that approval to every entity they interact with.

Additional options from the more menu include saving the Page to look at later, sharing it to your timeline, and reporting or blocking the Page. You can also use this menu to suggest edits to the Page admins or like the Page as one of your Pages (more on that in the "Using Facebook as Your Page" section).

>> **About:** On the left side of the screen, under the Home tab, is an About box with information about the purpose of the Page, links to any external websites, and the type of Page it is (such as Public Figure).

>> **Timeline and the share box:** The timeline is the heart of a Page — it's where the admins post updates and where fans leave posts and comments. Scroll down to see posts the Page has added. The share box is identifiable by its large Create Post button. Pages may or may not have a share box, although most choose to let their fans interact as though they were friends (the share box on Pages has fewer options than on your timeline — usually just Status, Photo/Video, and Check In).

>> **Left side:** Down the left side of the Page, under the About section, you may see some additional areas you can interact with. One is the Suggest Edits box. If a Page represents a bricks-and-mortar business, you can suggest that the business update its hours, add a website, or post a menu. In this way, fans can keep a Page accurate and more useful for future visitors. The other section we want to discuss is the Page Transparency box. Click this box to see more information about who runs the Page and where in the world the Page managers are located. You can also see when the Page was created and if it's running any advertisements. This information helps you to determine whether the Page is representing an honest business or entity or if it feels scammy. In general, it's a good idea to further research unverified Pages that seem dubious or don't have a presence outside Facebook — such as a website, an Etsy store, or a physical location — before you interact with them.

REMEMBER

Pages that don't officially represent something or someone — in the past, these were known as Community Pages — tend to cover a wider range of things, from basic activities to popular opinions. Often, the object represented by these pages is something that simply can't be owned by a corporation or an individual — things such as "cheese" or "I love dogs." This type of Page exists so that pretty much everything in the world has a way to be represented on Facebook. You can see the I Love Dogs Page in Figure 14-2.

Connecting and interacting with Pages

Wherever you go on Facebook, and in many places across the entire Internet, you'll see links and buttons and icons prompting you to like something. You can like photos, statuses, comments, articles, websites, videos . . . if it's online, you can probably like it.

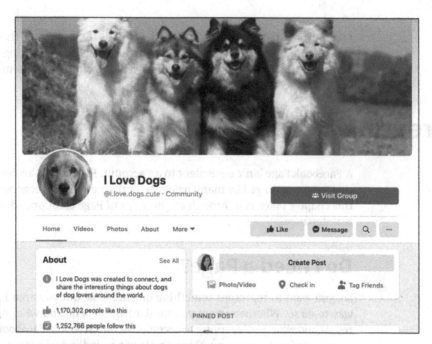

FIGURE 14-2:
Ode to man's
best friend.

You can also like Pages. When you like a Page, you can

>> **Subscribe to the page in your News Feed.** After you like a Page, you may begin seeing its posts in your News Feed. If you don't like what you see, you can remove that Page from your News Feed. To do so, go to the Page, click the three dots icon, and then click Follow Settings to adjust the frequency of posts or unfollow the Page (but you continue to like the Page).

>> **Get suggestions for other Pages.** When you like a Page, Facebook may begin suggesting other Pages for you to like in the Related Pages box on the left side of the Page. This feature is a great way to find Pages with useful and interesting content. For example, after you like the Page of a local coffee shop, you might see the Page for a local wine shop.

>> **See the Page on your own timeline.** Keep in mind that when you like a Page, it appears in the Likes section of your timeline, and may appear as a News Feed story to your friends.

So what does this all mean for you? Basically, when you like something, you start a relationship with it that can be as interactive or as hands-off as you want. Frequently, people like a lot of Pages simply as a signifier or badge on their timelines. However, just because you like the TV show *This is Us* doesn't mean you want to read episode recaps or watch interviews with the cast. And that's fine. On the other hand, if you like seeing those sorts of things or interacting with other fans on the show's Page, you can do that as well. It's a flexible system.

Creating a Facebook Page

A Facebook Page isn't equivalent to an account. Rather, it's an entity on Facebook that can be managed by many people with their own distinct accounts. The rest of this chapter takes you through all the steps of Page creation, administration, and maintenance.

Do I need a Page?

So, you want to represent something on Facebook, but you aren't sure of the best way to do so. Whether or not you need a Page depends on what you want to represent and what you're trying to accomplish on Facebook. Facebook Pages, at their most basic level, are for anything that's not an individual person and can even be useful for individuals who are celebrities or who have a public presence beyond their friends and family. This category includes small businesses, big businesses, bands, pets, charities, products, and much more.

Pages are best at promotion and distribution. If you want discussion and collaboration, groups might be a better fit. For example, if you're organizing a neighborhood fun run, you might create a Page to represent the event itself, post training schedules and weather updates, and generally let people know what's going on. At the same time, you and other organizers might create a group where you can discuss logistics. The Page is a public presence for everyone and the group is for the people running the show.

A NOTE ABOUT COVID-19

As we write this book, the Covid-19 pandemic is affecting the world and changing how we go about our daily lives. Facebook is attempting to do its part to help businesses weather the storm by giving Page admins the ability to add Covid-19 updates concerning any changes to their policies, business hours, or product availability. Whether or not Pages choose to add these updates, it's always a good idea to send a quick message to double-check things like closing times and delivery options.

So whether you have a small consulting business, are fundraising for a local organization, or are a member of a performance troupe, creating a Facebook Page can work for you. You *do* need to be an authorized representative of any larger entity (for example, you shouldn't create a Page for a local congressperson unless you're working for her). But assuming that part is all squared away, you're ready to find out how to create and manage a Page.

Creating your Page

Before you get started, we recommend that you read the Facebook Pages Terms at www.facebook.com/policies/pages_groups_events/. The terms clarify some of the expectations for owning a Page and who can create a Page for a business. One thing to note: You must have a personal Facebook account to create a Page and act as its admin. Your personal account will not be public information, so unless you broadcast the fact, your fans won't know which personal Facebook account is managing the Page.

There are also a few notes about who can see your content when you create a Page and age restrictions on your Page. We cover these topics throughout this chapter, but the terms provide a nice summary. If you violate these terms, your Page may be disabled, which could negatively affect your business. On the same note, if you use a personal timeline to do the work of a Page (as we describe here), your Facebook account will almost certainly be disabled for violating the Terms of Service (which you can read at www.facebook.com/terms.php).

REMEMBER

Pages can have multiple admins. If you plan to have other people managing the Page you're creating, they can do so from their own accounts. There's no reason to share the email address or password with anyone. Use your real email address and birth date; don't create a fake persona just for the Page. Your information won't be revealed to anyone else, and it makes your future interactions on Facebook much easier. To create your own Page, follow these steps:

1. **Log in to Facebook with your username and password.**

2. **Click the + icon in the top bar and choose Page from the drop-down menu.**

   The Create a Page page appears, as shown in Figure 14-3. This page should look familiar because it's the same interface used to create events and groups.

3. **Enter the name of your Page.**

   Use the exact name of your business, just as you need to use your real name on your Facebook account. For example, if you are creating a Page to represent Anthony's Pizza, that should be the name of your Page, not *Anthony's Pizza Official Facebook Page.*

FIGURE 14-3:
Fill in your
blank Page.

4. **Choose a category for your Page.**

   Type in the text box and Facebook will attempt to autocomplete the category.
   If you type Shop, you will get results such as Shopping and Retail, Shopping
   Mall, and Coffee Shop. You can choose up to three categories per Page.

5. **Add a description.**

   Include as much pertinent information as possible. What or who is the Page
   for, how do fans interact and contact admins, and what services are provided?

6. **Click the Create Page button.**

   Remember, when you click Create Page you are agreeing to the Facebook Page
   terms.

Getting started

In this section, we go over some of the things Facebook prompts you to do when
setting up your Page. This example is categorized as a Food and Beverage service,
so your own Page may look slightly different.

Step 1: Add a profile picture

The first step is to get your Page's profile picture in place. Click Add Profile Picture
to search for an image from your computer's hard drive. Click Open or Choose to
select the photo and upload it.

When the photo is successfully uploaded, it appears in the Page preview, as shown
in Figure 14-4. You can always restart the adding photo process by clicking the
trash can icon in the upper-right corner of the photo to remove it.

FIGURE 14-4:
Set up your Page.

FIGURE 14-4:
Set up your Page.

TIP

Profile pictures are 170 x 170 pixels; anything larger will be cropped to fit in a circle. If you decide you want to upload a different photo after you've published your Page, you can edit your profile photo by clicking the camera icon and choosing Edit Profile Picture, where you'll be able to upload a photo and adjust the position so it's centered in the circle.

Step 2: Add a cover photo

You want to make a good first impression with your Page, and one of the easiest ways to do that is to choose a striking cover photo. To add it, click the Add Cover Photo button and go through the same process of uploading a photo from your hard drive. You can adjust this photo by clicking and dragging to reposition. When you're sure the image is positioned properly, click the blue Save button.

TIP

A video may help you say more about your business than a photo. Cover photo videos are short videos meant to catch the eye of anyone looking at your Page. Videos must be 20 to 90 seconds long and a minimum of 820 x 312 pixels. To add a cover video, click the Edit button in the cover photo and select Choose from Videos in the drop-down menu. You'll be able to choose from a list of qualified videos that you've previously uploaded to your Page. You will not be able to upload just any old video — it must have been added to your Page first.

REMEMBER

Unless we specify otherwise, these instructions are meant for using Facebook on a computer. Creating a Page using your mobile phone is a slightly different process but designed to be simple and intuitive.

Step 3: Set up your page for success

After adding your profile picture and cover photo, you Page is public and ready to go. Facebook has a handy list of everything you still need to do to finish your Page. The list is on the left side of the Page under the heading Set Your Page Up for Success, as shown in Figure 14-5. We recommend going through the list one by one. If applicable, add a website, business hours, phone number, and most importantly, a call-to-action button, which we discuss in Step 7.

Step 4: Create a unique address

Click Create @Username next to the profile picture to choose a unique moniker for your Page. This also gives your Page a vanity URL (also known as a web address) that's easy to

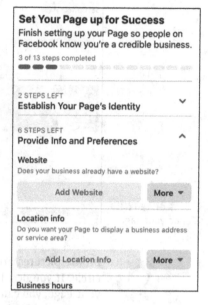

Set Your Page up for Success
Finish setting up your Page so people on Facebook know you're a credible business.

3 of 13 steps completed

2 STEPS LEFT
Establish Your Page's Identity ⌄

6 STEPS LEFT
Provide Info and Preferences ⌃

Website
Does your business already have a website?

Add Website | More ▾

Location info
Do you want your Page to display a business address or service area?

Add Location Info | More ▾

Business hours

FIGURE 14-5:
Don't forget your Page to-do list.

direct people to. If a local coffee shop decides to pick @seattlecoffeeshop as their username, their URL will be `http://facebook.com/seattlecoffeeshop`. See Figure 14-6 for the username interface.

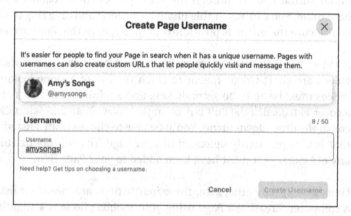

Create Page Username ✕

It's easier for people to find your Page in search when it has a unique username. Pages with usernames can also create custom URLs that let people quickly visit and message them.

Amy's Songs
@amysongs

Username   8 / 50

Username
amysongs|

Need help? Get tips on choosing a username.

Cancel | Create Username

FIGURE 14-6:
Choose wisely.

WARNING

The option to create a username may be temporarily unavailable if you've created multiple Pages in a short period of time. Usernames may also be revoked due to Page inactivity, freeing them up for an active Page to use.

Step 5: Like yourself

Now that your Page is up and running, you can be the first person to like it! Head to the three dots icon and click Like to become your first fan.

Step 6: Invite your friends

One of the first steps in starting your Facebook account is to get some friends. This step is even more important for Pages: Without likes, your updates and information won't reach anyone. A good way to start getting likes is to suggest your Page to your own friends. Look for the Invite Friends button on the left side of your Page or choose Invite Friends from the three dots icon. Enter the names of your friends in the text box and select the box next to their names when it appears. Click Send Invites when you're finished and your friends will receive a notification that you invited them to like this Page.

Step 7: Add a call-to-action button

You can still do plenty of things to customize your Page at this point, but one we highly recommend is adding a large button under your cover photo with a call to action. Click + Add a Button and choose the button that best fits your Page. For example, a clothing store may choose the Shop Now button while an HVAC company may find a Get Quote button more useful. After you choose a button, it is immediately added to your Page. You can edit or switch out your button any time by clicking the button (it will say Edit <*button name*>) and choosing Edit Button from the drop-down menu that appears.

Step 8: View as visitor

Our final piece of advice is to step outside your role as the Page's creator and click the View as Visitor button to see your Page from the viewpoint of someone who happens upon it. Is it easy to find opening times? Is the cover photo eye-catching? Revisit this tool from time to time to stay informed.

Sharing as a Page

After completing the preceding steps, you'll see a fairly empty Page timeline. You can add a bunch of content immediately or, as with your personal timeline, fill it up over time. Your Page will be visible on Facebook right away, so after you create it, be ready to use it on a regular basis so that it remains active and interesting to people.

The share box

If you have a personal timeline, your Page's timeline will look familiar; it's the virtual scrapbook of posts you've added to Facebook. Your Page also needs a virtual scrapbook.

If you're a Page, the same things are true: People are going to want to hear from you and learn about you, and the place they go to do so is your timeline. Any content you post to your timeline also shows up in your fans' News Feeds. In other words, the timeline is an important place to represent yourself honestly and engagingly through continual updates.

REMEMBER

When we talk about *you* in this context, we mean *your Page*. So if you're a coffee shop, people want to know about your seasonal drinks. If you're a band, people want to know about your shows. If you're a dog, people want to know what sort of trouble you've gotten into recently.

As a Page admin, the most important part of the timeline for you to understand is the share box. The *share box*, shown in Figure 14-7, is where you and your fans create the posts that populate your Page.

The most basic post you can make is a *status update*, a short message letting people know what's going on, what you're up to, what you're

FIGURE 14-7:
Use the share box to send posts out to fans.

thinking about, and so on. Just like status updates from your personal timeline, updates from your Page can be done quickly and easily, or with more options.

The basic steps follow:

1. **Click the Create Post button.**

2. **Type your update where it says Write Something to <*Page name*>.**

3. **Click the blue Post button.**

   The post now appears in your timeline and will potentially appear in the News Feeds of people who have liked your Page.

Add photos and videos to your post

Post a quick photo or video — for example, a company posting a photo from its holiday party, by clicking the photo icon at the bottom of the share box. An interface opens for exploring your computer's hard drive. Click the photo(s) or video(s) you want to add and then click Choose or Open. You return to the share box, and your selection appears at the bottom of the post. Photo posts are straightforward. You can add text, click Post, and be done. Video posts are a little more involved, as shown in Figure 14-8.

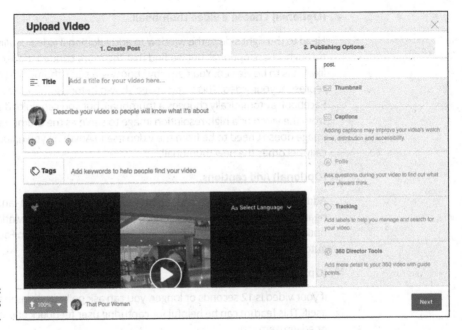

FIGURE 14-8: Video options galore.

To post a video, follow these steps:

1. **Select the Photo/Video icon at the bottom of the share box.**

2. **Select the video you want to upload from your computer's hard drive and click Open or add it to your post.**

   You're taken to the upload video window and prompted to fill out the Video Details section.

3. **Add a title.**

   We advise that you use a few attention-grabbing words, such as *Backflip into the pool*.

4. Add a description.

Tell people what your video is about so they're more likely to watch it.

5. Add tags.

Add tags to make your video more searchable.

6. Select a language.

Click the Select Language button in the top-right corner of the video preview, under the tags section, to change the language in which the video will appear.

7. (Optional) Choose a video thumbnail.

Head to the right side of the window to add additional video details. Steps 7 through 11 are optional, but we find that the more curated a video is, the more likely it is to be viewed. You begin by choosing a static thumbnail photo to represent your video. This is the image shown before someone clicks play. Facebook automatically chooses a thumbnail, but you can upload a still frame from the video or a high-resolution image to serve as the thumbnail. (This image doesn't need to be from the video itself.) Anything you upload automatically becomes the new thumbnail.

8. (Optional) Add captions.

Adding captions makes your video more accessible to the deaf- and hearing-impaired communities. Check out the left side of the window for options such as writing your own captions instead of relying on Facebook's auto-generating captions.

9. (Optional) Add polls.

If your video is 12 seconds or longer, you can ask poll questions in the video itself. This feature can be helpful for capturing user feedback about products or advertisements.

10. (Optional) Add tracking labels.

Tracking labels, which are for Page admins only, help you locate the video in your Page. This feature is useful for Pages that frequently upload videos.

11. (Optional) Use 360 tracking tools.

This cool feature renders your video in 360 degrees, meaning you can direct the audience to specific view points or details as the video plays. After you set the guide points, the video will automatically pan to them during playback.

12. Click Next.

13. Choose your publishing options.

You can publish immediately or schedule the video's launch. You can also display the video just on your Page or add it to Facebook's video-on-demand service, Facebook Watch. You can also make it embeddable on other websites outside Facebook and turn off reactions, comments, and shares on the video.

14. Click the blue Publish button.

Your video now appears on your Page and will show up in your fans' News Feeds.

REMEMBER

In general, you always have more options for customizing your photos and videos by using the Facebook mobile app.

Add feeling or activity info to your post

You can add information about what you (not you personally, but you the Page) are doing by clicking the smiley face icon at the bottom of the share box. Doing so opens a list of various activities, and you can choose from any number of emotions your Page might be feeling or activities your Page might be doing. When you choose to add activity info, that info gets appended to the beginning of your post along with an emoji (or icon) representing that activity or feeling. You can see what this looks like in Figure 14-9.

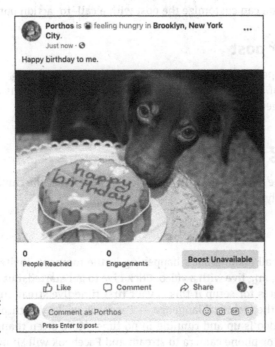

FIGURE 14-9:
Share your
feelings.

Add location info to your post

You can add information about where you are when you post something by clicking the three dots icon at the bottom of the share box and choosing Check In. A text box appears so you can type your location. Facebook attempts to autocomplete as you type; when you see the place you're looking for, click it.

Location info gets appended to the beginning of your post. You can see an example of this in Figure 14-9.

REMEMBER

Some share box options exist for only certain types of Pages. For example, a charity Page can start a fundraiser right in the share box by clicking Raise Money, whereas a musician's Page can start a watch party for fans.

Publishing tools

Many Pages find it useful to coordinate their posts with events or real-world happenings. You might not want to promote your holiday sale until a certain time or date, or you might want a specific post to appear on Valentine's Day, but you don't want to be sitting at your computer at either of these times.

At the bottom of the share box is a blue Publishing Tools link and if you click it, you're taken to the Publishing Tools page. From this page, you can schedule a specific date and time to publish the post, you can cross post it to your Instagram account, and you can customize the post with a call-to-action button.

Boost your post

Boosting your post is a form of paid promotion. In other words, it's a Facebook-specific form of advertising. Clicking the Boost Post slider opens an interface for selecting a target audience and setting an ad budget.

Creating specialty posts

Under the share box is a short list of specialty posts that you can create. We'll go over each briefly here.

Live post

Nowadays, it's all about what's happening in the moment. Invite your fans into your world by going live with a video. Click Live to get started, but be sure to budget a few minutes for setup if this is your first time live streaming. Don't be put off by some of the technical language — Facebook walks you through everything so that your video is up and running in no time. More often than not, you'll use your computer or phone camera to stream and Facebook will show you a preview of what that looks like in the bottom-right corner of the Live setup page. Fill out the necessary info, such as video title and description, in the left sidebar, and then click the Go Live button to start your video. Check out the top of the page for options such as adding polls or questions in real time. Note in the left sidebar that you can also schedule a future live video event.

Events post

If you ever host an event for your business, you'll get a ton of value from using Facebook Events. Stores create events for their big sales, comedians create events for their shows, and clubs create events for their theme nights. Click Event, and then choose whether your event is online or in person as well as if it's free or paid. Continue adding information until you're ready to publish the event. Head over to the event's page to invite your friends. Events you create for your Page are public by default, meaning anyone can see and join the event. Figure 14-10 shows a Page's event.

FIGURE 14-10:
Add an event to
your Page's
timeline.

If you haven't used Facebook Events before, see Chapter 13 to learn about creating and managing your own events.

Offer post

Depending on your Page type, you might have the ability to post offers for fans. Click Offer to go to the Create New Offer page, where you will add information such as discount type, where the offer is available, promo codes, and any terms and conditions. This feature is particularly useful for retail businesses.

Job post

There has been a big push on Facebook lately to post jobs. You can find job listings in groups, in News Feed, and now on Pages. And why not? Job hunting is often about who you know and getting your foot in the door, so what better place to look than where all your friends are. If your business is hiring, you can post a listing by clicking Job and filling out the Create Job page, where you add details such as position title, location, duties, and salary.

Ad post

The final option is creating an ad for your Page. We could fill an entire book with everything you need to know about Facebook ads and how to best monetize your business, but for now we'll just say that creating an ad is simple and Facebook excels at targeting ads to those who will be the most interested in the content. So for example, if you own a local coffee shop, you could ensure that your ads are seen only by fellow community members and that you're spending money on attracting actual customers instead of someone across the country who will never visit your shop.

Using Facebook as Your Page

Part of using Facebook as a Page is being interactive. Pages aren't just ways for you to distribute information to people who like your business or band or book. Pages are a way for you to engage in conversation and be human (or as human as a Page can be) with people. Part of that process is liking and commenting on behalf of your Page.

Liking, commenting on, and sharing posts

As a Page admin, when you're looking at posts from other Pages on Facebook, you have the opportunity to like and comment on many of them *as* your Page. When you see a public post from another Page, look at the bottom-right corner of the post, where a tiny thumbnail of your (personal) profile picture appears next to a down arrow. Clicking that arrow opens a window, shown in Figure 14-11, for selecting who you want to comment as.

Select the Page you would like to use to comment or like, and then click any of the links at the bottom of the post to like, comment, or share that link on your Page.

When you like, comment, or share as a Page, changing who is commenting works for only that particular post. In other words, if there are ten articles you want to comment on as your Page, you will need to switch from your personal account to your Page, as described, each of those ten times.

FIGURE 14-11:
Like or comment as your Page.

Liking other Pages

You can like other Pages as your Page. To do so, check out the following steps. For clarity, we reference your Page as LocalCoffeeShop and the Page you want to like as MajorCoffeeBrand.

1. **Go to the Page you want to like (MajorCoffeeBrand).**

2. **Click the three dots icon under the cover photo.**

   A list of options opens.

3. **Click Like as Your Page.**

   A confirmation window appears where you can choose the Page you want to use to like MajorCoffeeBrand.

4. **Use the drop-down menu to choose LocalCoffeeShop.**

5. **Click Submit.**

Liking other Pages is a great way to be involved with the other entities your Page interacts with in real life, and also a way to get ideas for what to post about on your own Page. For example, if you're a restaurant and you like the Pages of other restaurants nearby, you might learn about local events that you also want to promote. There are so many possibilities for using your Page over time; in this chapter we just give you the basics. Other Pages can teach you a lot.

Managing a Page

The basics of using a Page are fairly simple: Pages are timelines for non-people or people in the public eye; use them to share and interact with your customers and fans. Up to this point, this chapter has focused on those basics.

After you have the basics down, however, it's worth noting that Pages is an incredibly powerful and diverse product. Pages can be created to represent the family dog or the Smithsonian Museum. The needs for those two entities are pretty different, and yet Pages accommodates them both. In this section, we go over many of the settings and tools that Pages offers beyond simple sharing. Some tools might seem superfluous. Some settings might not apply. We might miss certain settings that apply to only certain Page types. If you have a question that we don't answer here, check out www.facebook.com/business/pages. Facebook provides a lot of material for Page owners to explore and learn.

When you look at your Page, note the left sidebar. Pictured in Figure 14-12, these tabs are visible to only the Page's admin. People who like the Page don't see these tabs.

The first entry, Home, simply brings you back to your Page's timeline. The others are where you find options that help you get the most from your Page.

FIGURE 14-12:
Page management sidebar.

News Feed

Facebook recently added the News Feed feature, which provides you with a separate News Feed for your Page. You'll see content from any Pages or public figures you've followed as your Page as well as from related Pages that Facebook thinks you may be interested in. Facebook also likes to suggest Pages for you to follow. For example, if you're a coffee shop, you might be prompted to follow the coffee shop on the other side of town. You can access your Page's News Feed only when you're interacting as your Page.

Inbox

Much like your personal Facebook account, Pages have an inbox where you (and other admins) can respond to messages from people who encounter your Page. You receive a notification whenever you have a new message in your Page inbox. People don't have to like your Page to message it, and the sorts of messages you receive may vary. Figure 14-13 shows a sample inbox, with an open message.

FIGURE 14-13:
A Page inbox.

At first glance, this inbox looks similar to your personal message center. Your message list is displayed on the left side of the page, with the newest messages at the top of the list. Unread messages are in bold. The open message is displayed in the center of the page. You read a message thread from top to bottom (scroll up to see the oldest messages) and you can type your reply at the bottom of the thread.

Unlike your personal inbox, though, the right side of the page displays any publicly available info about the person with whom you are messaging. This information gives you a sense of the person you're talking to — where the person is from, where he or she works, all the sorts of things that might be important to learn if someone is trying to contact you to make a professional connection.

Depending on what your Page represents, you might get a lot of customer support questions in your messages. In those cases, it might be less important to know the biographical details about your customer, but it never hurts to put a name with a face when talking to someone.

You can use the search box on the top of your inbox to search through your Page's messages by keyword or name.

TIP

If you have an Instagram account in addition to a Facebook Page for your business, it pays to connect the two and manage messages from both channels in one central inbox. You'll see lots of prompts to make that connection. Just ignore them if you don't have an Instagram account or if you want to keep Instagram separate.

Message options

When you're looking at the list of messages down the middle of the page you have a few options for keeping track of them. Click any individual thread to see a list of icons above the public information about that person.

» **Move to Done:** Click the check mark icon to move this message thread to your Done folder. This keeps your inbox organized and limited to active message threads. If you need to access your Done folder, click the down arrow next to Main on top of the list of messages and choose Done.

» **Delete Conversation:** If you want to get your inbox to zero messages, you can delete messages permanently by clicking the trash can icon. You'll be prompted to confirm that you indeed want to delete the message forever.

» **Move to Spam:** If you receive messages that are spam, click the exclamation point icon to remove them from your inbox. Messages you mark as spam are moved to your Spam folder, which you can access by clicking the same down arrow you use to get to your Done folder.

» **Mark as Unread:** You can choose to mark messages as read or unread by clicking the envelope icon. We usually mark messages as unread when we intend to reread them later. Marking a message as unread bolds the text to help it stand out from the list of messages.

» **Mark as Follow Up:** Many people leave messages they need to follow up on as unread, but Facebook does you one better here and gives you a separate designation for messages that need further action. Simply click the star icon to mark a message you need to follow up on.

Replying to messages

The basics for replying to messages as a Page are pretty much the same as those for any email, text, or instant message service. Type your response and click Send.

Facebook also provides the ability to send *saved replies,* which are common replies that you use to more quickly respond to people. Are people always writing to ask what the delivery window is for your pizza shop? Create a saved reply with your delivery hours. Then, anytime someone writes asking that question, you can respond in just a few clicks; no typing necessary. Click the speech bubble icon to open the Saved Replies window, shown in Figure 14-14.

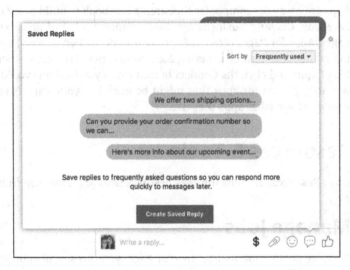

FIGURE 14-14:
Create your saved
reply here.

Saved replies are designated by titles (which the person you're messaging doesn't see). If you have a lot of saved replies, you can search through the options using the search box at the top of the Saved Replies window. Click the reply you want to send, and the reply box is filled with the text from the saved reply. Then click Send.

Click the Create New Reply button to get started. This opens a window for entering the title and text of the saved reply. Click to personalize the saved reply with things such as the name of the admin who is replying or the name of the person who originally sent the message. You can also add images from your computer's hard drive. Click Save when you've finished creating the reply.

Automated responses

Automated responses require even less effort on your part than saved replies. These are replies that are sent immediately after receiving an incoming message, without any need for someone to look at the message first. Click Automated Responses from the left sidebar to see your options. For each type of message, click the blue Edit button to create or manage the message. Be sure to click the Save button up top after making any changes. You can set up an instant reply that everyone who writes in will receive, or an automated away message when you're on vacation and don't want to think about business. You can even avoid having to create saved replies for frequently asked questions such as location and contact information. Just think of all the time automated responses will free up!

Business App Store

Apps are a way for companies that aren't Facebook to build programs that work on Facebook. Clicking Business App Store displays a list of apps built for Facebook Pages. Apps for Pages often add certain functionalities to your Page, such as booking an appointment or ordering food from a delivery service. Scroll through the list of apps and click the Connect button when you find one you'd like to add. One way to get ideas for apps that might be useful to your Page is to look at similar Pages and see what apps they use.

Resources & Tools

Go to Resources & Tools for Page best practices and guidance from Facebook.

Manage Jobs

If you're posting job listings to your Page, you can see all submitted applications for a specific position by clicking Manage Jobs.

Notifications

Whenever anything happens to your Page such as someone liking it, liking a post your Page made, commenting on a post, or sharing a post, a notification is created. All the notifications about your Page are collected on the Notifications section, shown in Figure 14-15.

Notifications may or may not be something that winds up mattering to you as a Page owner. In general, evidence of people interacting with your Page and the content you post is considered to be a positive sign of engagement, so it's worth noting if the number of notifications suddenly changes a lot or a little. Unread notifications have black text and a blue dot, and read notifications have dimmed text.

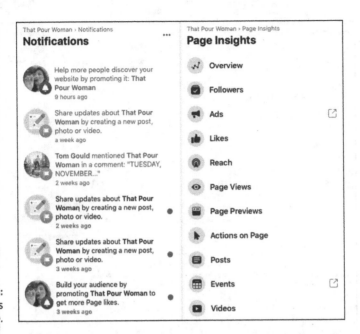

That Pour Woman › Notifications

Notifications

Help more people discover your website by promoting it: **That Pour Woman**
9 hours ago

Share updates about **That Pour Woman** by creating a new post, photo or video.
a week ago

Tom Gould mentioned That Pour Woman in a comment: "TUESDAY, NOVEMBER..."
2 weeks ago

Share updates about **That Pour Woman** by creating a new post, photo or video.
2 weeks ago

Share updates about **That Pour Woman** by creating a new post, photo or video.
3 weeks ago

Build your audience by promoting **That Pour Woman** to get more Page likes.
3 weeks ago

That Pour Woman › Page Insights

Page Insights

Overview

Followers

Ads

Likes

Reach

Page Views

Page Previews

Actions on Page

Posts

Events

Videos

FIGURE 14-15:
Notifications go here.

In addition to notifications of actions, the Notifications section shows you a record of activity and information requests. *Activity* includes things such as people mentioning (or tagging) your Page in a post. *Information requests* show up when someone comes to your Page and wants to find out something that isn't there. For example, if you didn't add a website to your Page when you set it up, people can click a link to ask for your website address. Those requests come to this portion of your Page's Notifications section.

You can control how you receive notifications from the Settings section. We go over that later on in this chapter, in the section "Notifications setting."

Insights: Finding out who is using your Page

The Insights section contains all anonymized data that Facebook can give you about who is looking at your Page, what they liked or didn't like, how they've been interacting with your Page, and how engagement has changed over time. You can see a sample Page Insights Overview in Figure 14-16.

Similar to many of your Page's management sections, a left sidebar lets you dig into different statistics: Likes, Reach, Page Views, Posts, Videos, and Followers. By default, you land on Overview which gives you, well, an overview of all those different stats.

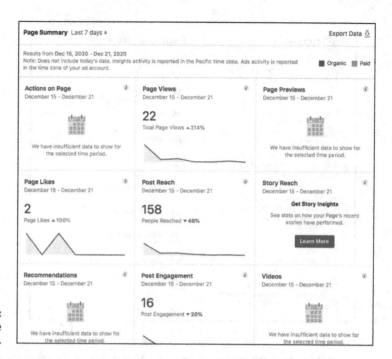

FIGURE 14-16:
View your Page
insights here.

On the Overview page, look at the Page Summary section to see the date range of the data. Below are ten boxes, each with some numbers and graphs. We cover the three data points you'll be looking at the most, but you can get pretty granular by clicking for a detailed view of each data point from the left sidebar.

>> **Page Likes:** Likes are the most basic metric for measuring how your Page is doing. Likes aren't everything; if you're trying to draw attention to your Page, someone clicking Like and then never visiting your Page again or reading anything you post isn't so great. But Page Likes is still an important baseline number to pay attention to. The Page Likes box shows you how many likes you have, and how much the number has changed day to day or for a time period you set. If you notice a sudden uptick or downtick in likes, you might want to think about anything your Page has done or posted in the last week that might have mattered to people.

>> **Post Reach:** Reach shows you how many people have seen posts you've made in the last week. Hover your cursor over the graph to see how many of the reaches were *organic,* meaning people found the Page on their own versus from a promotion or ad campaign.

>> **Page Engagement:** Post Engagements keeps track of how many people have liked, commented, shared, or clicked one of your posts in the last week.

None of these numbers, alone, really matter. But together, they paint a picture of what your customers and fans are doing, and whether you're truly interacting with them or simply shouting into the void. The way these numbers trend over time helps you learn more about your customers and fans and how they want you to engage with them.

Below the ten boxes are insights about your five most recent posts. Each post gets a row with a link to the post if you want to view it. You can see what type of post it is (link, picture, video, and so on), how it was targeted (everyone or a specific demographic), the reach of that post, and the engagement with that post. Again, you're not looking to achieve any one particular goal here, but you can learn a great deal. You might see that you have more engagement with photo posts, or greater reach when you target a post by demographic. Because managing a Page is an ongoing process, these insights inform decisions about how you run your Page and share content with your fans.

Below this you can click Add Pages to monitor up to five similar Pages and their likes. So, for example, the local coffee shop can keep an eye on its competitors and see if they're doing anything to drive traffic that the local coffee shop can implement as well, such as offering a customer loyalty card.

TIP

If you consider yourself a content creator, you may want to go one step further and check out the insights on Facebook's Creator Studio, which you can see at https://business.facebook.com/creatorstudio/home. From here you can upload and manage content, monetize with paid online events, and use creative tools such as the live dashboard for streaming games, which we describe a little more in Chapter 15.

Publishing tools

The Publishing Tools section gives you the ability to quickly review your posts — ones you've already made, ones you've drafted, and ones you've scheduled. As shown in Figure 14-17, the Publishing Tools section shows you a list of your posts, along with some data about those posts.

The left side of the Publishing Tools section categorizes your posts so you can quickly get to the type of post you want to review. For example, if you're logging in to see what posts other admins have scheduled for the upcoming week, you can click Scheduled Posts to view only those. You can also use this sidebar to jump to your video library, pending orders, or events you've planned.

When you're looking at Published Posts, you can see a thumbnail of whatever image accompanied the post and the title of it (if it has one). You can then view your reach for that post (the number of people who saw the post). Because News

Feed (the main way your posts are distributed) works on an algorithm to ensure each person sees what is most likely to be interesting to him or her, not all the people who have liked your Page will see your post. You can also see how many people engaged with your posts, which often means people who reacted or commented. Finally on the right side of the screen, you can see the date your post was published.

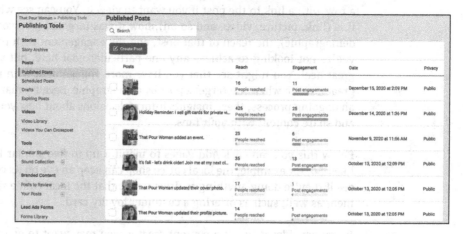

Ad Center

Below Publishing Tools is Ad Center. Facebook makes it easy to create an ad to help promote your Page and potentially convert people who see the ad into new customers. Click the big blue Create Ad button to get started.

Page Quality

Anyone who comes across your Page can report it to Facebook for offensive, abusive, or fraudulent content. If Facebook reviews the report and takes action against your Page, you'll see those violations listed in the Page Quality section. Depending on what Facebook's review uncovers, they may also place restrictions on your Page, such as curbing your ability to post for a period of time.

Edit Page Info

Head to the Edit Page Info section to quickly update any information about your Page.

Page settings

As we mentioned, Pages is a powerful and in many ways flexible system that accommodates virtually every type of entity. For a system to be that flexible, it needs to be highly customizable. And nowhere is the ability to customize more apparent than on the Page Settings section (see Figure 14-18).

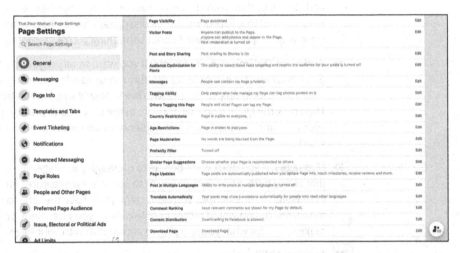

FIGURE 14-18:
Settings, so many settings.

In this section, we go through each setting and try to include examples of how some of the more obscure settings might be utilized. Remember, depending on your Page and how you use it, some of these settings might not be relevant. Certain settings will seem especially irrelevant if you're just starting out with your Page and don't have many fans. After your Page picks up in popularity, however, revisit some of the settings that didn't apply when you were starting out.

General setting

General settings cover how your Page posts and interacts with other people and Pages on Facebook. To open the settings related to these categories, click Edit to the right of each setting. When you adjust a setting, remember to click Save Changes.

>> **Page Visibility:** This option controls whether or not your page is published (publicly visible). If you want to edit your Page without people seeing the edits, you can opt to Unpublish the Page for a period of time while you tinker and make changes. Other Page admins will still be able to see and access the Page.

>> **Visitor Posts:** By default, people can post to your Page as they would post on a friend's timeline. You can disable this option, allow people to be able to post text but not photos and videos, or review all posts before they appear on your Page. Posts that need to be reviewed show up in the Page's activity log, which is the last section on the left sidebar.

>> **Post and Story Sharing:** This setting allows fans to share your Page's posts and stories to their own stories on Facebook. You can disable this option, but keep in mind that Facebook stories disappear after 24 hours and can be a great way to spread the word about something you're promoting.

>> **Audience Optimization for Posts:** If you want to target your post to certain demographics (for example, a post about your band's upcoming show in New York shown only to people who live in New York), turn on this setting. Then click the globe icon that appears in the share box's privacy menu for your post and choose Restricted Audience to add parameters for age and location.

>> **Messages:** If you don't want people to message your Page, you can turn off that ability here. You might not want to use message features for any number of reasons — for example, you don't think you'll have the time to respond to messages, or you would rather people contact you elsewhere, or it doesn't make sense for what you're trying to accomplish with your Page.

>> **Tagging Ability:** Allow everyone to tag photos and videos you post, or let only you and other admins tag photos and videos posted by your Page.

>> **Others Tagging This Page:** Select this option to prevent people from tagging your Page in their own timeline posts. You can also turn on a setting to let people tag your Page in an Instagram story that they cross post to Facebook.

>> **Country Restrictions:** By default, everyone the world over can find and view your Page. You can hide it from certain countries or allow only certain countries to see it.

>> **Age Restrictions:** By default, anyone of any age (well, older than 13, which is the youngest age at which you can use Facebook) can see your Page. You can restrict your Page so that people under 17, 18, 19, or 21 can't see it. This option is most relevant if you're creating a Page to represent a brand of alcohol or tobacco. (An Alcohol-Related setting automatically adjusts the age restrictions for your Page.)

>> **Page Moderation:** Create a list of words that, if detected in a post, block the post from being posted.

>> **Profanity Filter:** Turn on Facebook's profanity filter, which has two settings, Medium and Strong. If someone tries to type a post with a profanity, that person will be unable to add the post to your Page.

>> **Similar Page Suggestions:** Control whether or not your Page will appear in Facebook's automatically generated Page Suggestions. By default, your Page appears in Page Suggestions.

>> **Page Updates:** A post is automatically published on your behalf if you update the Page buttons, description, contact information, or website.

>> **Post in Multiple Languages:** Allow other Page admins to post in multiple languages.

>> **Translate Automatically:** People who use Facebook in another language will see your posts automatically translated into their language. Note that this feature isn't available to every single user.

>> **Comment Ranking:** If your Page gets lots of comments and replies to every post you make, you may want to turn on comment ranking. This setting attempts to identify the most relevant comments (as determined by people liking and replying to that comment) and place them at the top of the comments section. If you leave this option off, the most recent comments are displayed first.

>> **Content Distribution:** Give users in some countries the option to download videos you've added to your Page. If you enable this option, rest assured that videos can't be downloaded to mobile phones in that country. Select this box to prohibit anyone from downloading your videos onto their computer.

>> **Download Page:** Click the Download Page link to receive a copy of your Page posts, videos, and Page Info. An email alerts you that the file is ready to download. This feature can be life-saving if content is accidentally deleted.

>> **Merge Pages:** If you wind up with more than one Page for the same thing (it happens), you can merge the two so that it's easier to reach everyone who has liked either Page.

>> **Remove Page:** If you want to delete your Page from Facebook, follow the prompts in this setting. You have 14 days to change your mind and reactivate your Page. No one can find your Page during this grace period. At the end of the 14 days, you'll be asked if you still want to delete your Page. If you proceed, the deletion is final and can't be reversed.

Messaging setting

Assuming you have kept your Page's capability to receive messages, you can adjust your settings in the Messaging section. We went over a few of these settings when we talked about Page inboxes. However, there are new options here:

>> **Your Messenger URL:** This web address is specific to your Page. You can post it anywhere to direct people straight to your Page's Messenger interface.

>> **Add Messenger to Your Website:** If you have a website for your business, you can add a Facebook Messenger plug-in so that people who reach out via your website can continue the conversation on Facebook. Click Get Started and follow the instructions to add the plug-in code to your website.

Page Info setting

The Page Info setting is the same as the Edit Page Info section in the left sidebar. We cover Edit Page Info a few paragraphs back.

Templates and Tabs setting

This section is all about customizing your Page to showcase pertinent information to visitors. For example, maybe an Events tab isn't important to your business, but a Reviews tab is. You can place the Reviews tab front and center and remove the Events tab completely:

>> **Template:** By default, your Page uses the standard template, but if you click the Edit button next to this setting, you can choose from other prefab templates, including ones for venues, restaurants and cafes, and video creators.

>> **Tabs:** The rest of this page is devoted to the tabs on your Page, such as Events or Reviews. Click the slider to remove that tab from your Page. Some tabs, such as About, are required and can't be removed. You can also drag and drop tab names to reorder how they appear on your Page.

Event Ticketing setting

If you use Eventbrite to manage your events, you can connect your account to your Facebook Page and manage ticketing using the Event Ticketing setting.

Notifications setting

Notifications always appear in your Notifications section for your Page, but you can also choose if they go into your Facebook notifications (where notifications for your personal account also appear), your email, or via text message on your phone by clicking the Allow Notifications slider to on.

If you want to see Notifications on Facebook but are feeling overwhelmed by the sheer amount, you can choose to receive a daily summary of notifications. You can also choose to turn off specific types of notifications (for example, you might want to be notified about new comments so you can reply right away, but you might not want to be notified whenever someone likes a post).

You can also turn off notifications about messages your Page receives. This setting can be helpful if most of the time you're using Facebook as yourself, and not your Page. Your Page's messages will always be available in the Inbox section.

Advanced Messaging setting

The Advanced Messaging setting is an advanced group of settings that you'll use if you build or use apps or bots in the Messenger section of your Page. You can learn more about your options for using these apps and bots at https://developers.facebook.com/docs/messenger-platform.

Page Roles setting

Page Roles represent the different responsibilities that people who manage Pages may have. Often, you aren't the only person who represents your Page. For example, if you have a small business, you may want all your employees to be able to post photos and announcements on behalf of your Page. Or you might need moderators to respond to people's comments and messages that you can't respond to yourself. Five distinct Page roles are available:

>> **Admin:** If you created a Page, you're automatically its admin, and you can control all aspects of your Page, specifically adding other people as admins or other roles, editing your Page's settings, editing the Page, posting from the Page, sending messages as the Page, responding to and deleting comments and posts as the Page, removing and banning people from the Page, creating ads, and viewing insights.

>> **Editor:** Editors have the same abilities as admins, but they *can't* edit all Page settings or assign Page roles to other people.

>> **Moderator:** Moderators have many of the same abilities as admins and editors, but they can't edit the Page or create posts as the Page.

>> **Advertiser:** Advertisers have the ability only to create ads, view insights, and see who created a post or comment. Lots of companies specialize in helping people manage their online advertising presence, which is why this is a unique position you can assign.

>> **Analysts:** Analysts can see your Page's insights and who created posts or comments.

As the admin, you can add someone to any of these roles from the Page Roles section. Type a name or email address in the text field under the Assign a New Page Role section, and then decide which role you would like that person to have from the drop-down menu to the right of the name field. Click the Add button to make it official. A notification is sent to the person and he or she can accept or decline the position. The person's profile picture thumbnail and the word *Pending* appear on the bottom of this page until the position is accepted. You can click Cancel Invitation to revoke any unaccepted position. If you want to remove employees who no longer work at the company, click the Edit button to the right of their name and then click the Remove link. You have to confirm you want to remove them from their page role. If you remove someone who was central to creating posts, you can transfer ownership of their scheduled posts to yourself or delete those posts.

People and Other Pages setting

The People and Other Pages section displays every person and Page who has liked your Page. Click the check box next to the person's name and then click the gear icon next to the search bar to do the following:

>> **Remove from Page Likes:** Removing people from Page likes means that person will no longer be shown as someone who likes your Page. It's basically a way to force someone to unlike your Page.

>> **Ban from Page:** If someone is being abusive or inappropriate on your Page, you can ban that person from interacting with your Page ever again.

Preferred Page Audience setting

Facebook recently removed the Preferred Page Audience section to simplify the settings. You may see a notice explaining the change.

Issue, Electoral, or Political Ads setting

If you're a political figure or work in politics, you must follow special rules regarding ads. The Issue, Electoral, or Political Ads setting gets you started on the authorization process to run these types of ads.

Ad Limits setting

Facebook places some limits on Pages that run a lot of ads to keep the ad experience positive for its users. Don't worry about the Ad Limits section unless you're operating a Page for a larger company that runs more than a few ads at a time.

Branded Content setting

Branded content is content you create to promote another company or product (it often resembles a regular piece of your content) in exchange for some kind of value to your company. As an example, if the popular website Buzzfeed decides to publish an article titled "10 Ways You Can Use Aluminum Foil That Will Blow Your Mind" and you notice that near the top it says "sponsored by Reynold's Wrap," that article is likely branded content.

From the Branded Content section, you can allow other Pages to tag your Page in a branded content post. (The default setting is off.) You may want to toggle this setting to on if you think these tags can increase your Page's presence and fan number. If you're a local coffee shop, for example, you may appreciate the boost that comes from being tagged by a major coffee brand.

After you turn this setting on, you can type the names of Approved Pages in the relevant field to ensure that only Pages you trust will be able to tag your Page. Conversely, you can also type the names of Pages that you want to block from tagging your Page in branded content.

You can also click Enable Tag to allow your Page to tag other Pages in your branded content posts. Before you start crafting branded content, be sure to read and follow Facebook's Branded Content Policy at `www.facebook.com/policies/ brandedcontent`.

Note that this section has its own left sidebar for keeping track of past partnerships, current projects, and insights to this particular type of content.

Instagram setting

From the Instagram section, you can link your business's Instagram account to its Facebook Page. After you link these accounts, you can manage ads, insights, messages, and comments from both sites in one place.

WhatsApp setting

WhatsApp is a free messaging service owned by Facebook that's particularly popular in Europe and Asia. Odds are, if you have friends living abroad, they've invited you to chat on WhatsApp to avoid fees associated with using their phone's default messaging app. If your business uses WhatsApp, you can add a WhatsApp button to your Facebook Page to make it easier for fans using WhatsApp to message you.

Featured setting

The Featured section is where you can choose to highlight certain Pages your Page has liked. A selection of five Pages your Page has liked always appears on your Page, so choosing to feature some merely allows you to specify which ones you most want people to know you like. This setting is a nice way to show love to small businesses you admire and want more people to know about.

Crossposting setting

Let's continue using the example of the local coffee shop. Perhaps they've part-
nered with a local artist to display her work on the walls of the shop. To promote
the partnership, the coffee shop films the canvasses being hung and posts the
video to the coffee shop's Page. In this scenario, the local artist may want to feature the same video on her own Page.

In the Crossposting setting, you list the Facebook URLs or names of Pages you
want to cross post videos on. You can add any Page you want, but that Page has to
add you as well before the ability to cross post on that Page is available. If the
other Page accepts your request to cross post, you'll see Crosspost as an option
from your Page's share box when you choose Photos/Videos. Note that the other
Page will be able to see the insights for that post. (They will not be able to see
insights for posts that aren't cross posted.)

Page Support Inbox setting

The Page Support Inbox is used for communicating with Facebook's Help Team
about your Page. If you ever wind up needing help from Facebook with your Page
(for things such as resetting your password or following up on reported material),
those messages can be found in the Page Support Inbox.

Page Management History setting

Under the Page Management History setting, you can see the history of actions
taken by those who manage the Page.

Activity Log setting

The Activity Log is a chronologically ordered list of every action your Page has
taken (such as posting a photo) as well as actions involving your Page (such as
posts added to your timeline). In the Activity Log setting, you can use the left
sidebar to filter by specific content type.

Chapter **15**

Using Facebook with Games, Websites, and Apps

I n this book, we spend a lot of time defining what Facebook is — it's a way to keep up with friends, it's a constantly updating newsletter about your friends, it's a way to share photos and posts. There's one more definition to add to the list: Facebook is a collection of information about you. It's not the cuddliest definition, but it's an important one to understand as you learn about Facebook Platform.

Facebook Platform is what Facebook uses to take your collection of information and (with your consent) make it available to other websites, game manufacturers, and app developers. What this means in practical, concrete terms is that you can choose to use your Facebook information to, for example, log in to a website such as Pinterest and share your pins with Facebook friends. You can also use your Facebook information to start playing games such as Words with Friends or Candy Crush Saga with your friends, both on Facebook and on your mobile phone.

The ways you'll see Facebook Platform integrating into your experience on Facebook may change over time. Right now, the emphasis is on games and ways to log in to other websites to then post content back to Facebook. No matter what forms Platform may take in the future, it's important to understand what you are sharing with apps. In this chapter, we explain the more common examples of how you'll wind up using apps, and how to manage your apps and your information over time.

Understanding What Apps Need

For the purposes of this chapter, when we talk about *apps*, we're talking about apps, games, and websites built by companies other than Facebook. But for the purposes of the next few sentences, we want to talk about an app that Facebook did build: Photos. As described in Chapter 11, Facebook Photos makes it easy for people to see photos you took and photos of you.

Photos works so well because Facebook knows some important information about who you are (your name and face) and who your friends are (people who want to see your photos). Facebook also knows that you're okay sharing your photos through News Feed, because, hey, you're using Facebook to share your photos. In other words, Facebook uses your information — who you are, who your friends are — to create a meaningful experience — sharing photos with your friends.

Some other apps, ones not built by Facebook, require the same things to work. They need to know who you are and who your friends are. Some may need to know when your birthday is or the type of music you like. An app such as Photos has your permission as soon as you start using Facebook, but websites, games, and apps need to obtain your permission before they can get this information. In this section we describe the information apps need and get when you use a game or website.

The basics

All apps need your basic information, also known as your public profile, if you want to use them. In other words, if you're not comfortable sharing this information with an app, you can't use it. Your *basic information* refers, in this case, to any information about you that is set to public. For everyone on Facebook, this information is as follows:

» Your name

» Your profile picture

» Your cover photo

>> Your gender

>> Your age range (for example, 21+ years old)

>> Networks (including professional and educational)

>> Language and country

>> Username (if you have vanity username associated with your Facebook URL, such as your full name)

>> Your user ID (the number associated with your Facebook account; everyone has a unique ID number)

>> Any other information you have shared publicly

In addition to this information, most games or apps, when you start using them, will request that you share your contact email address with them. The app can store that email address to contact you in the future. This allows you to establish a direct relationship with the app so that the developers can always get in touch with you, without Facebook acting as an intermediary.

Giving your email to an application means you can get email newsletters and other updates directly from the source without logging in to Facebook. If at any time you no longer want to share your email address with a certain application, you must unsubscribe from its email list through the app developer (in other words, through whichever company makes the app), not through Facebook.

The slightly less basics

In addition to basic information, apps might require information that is not publicly available. What information this includes depends on the app and what it does. Here are some examples of information an app might want to use:

>> Your list of friends (particularly to locate your friends who also use this app)

>> Things you like (books, music, movies, Pages, and so on)

>> Things your friends like (books, music, movies, Pages, and so on)

>> Location information (hometown, current city)

>> Your birthday

>> Your posts

Permission to act

In addition to all the types of information apps need from you, they also need permission to take certain actions, including the following:

>> **Posting to your News Feed, timeline, or group on your behalf:** For example, an app might create a post when you win a game. You can choose who can see these posts using the regular privacy menu, which is covered in Chapter 6. Later in the chapter, we explain how to revoke this permission on an app-by-app basis.

>> **Sending you notifications:** Notifications from games or apps are like the notifications you receive from Facebook about being tagged in a photo or invited to an event. Games or apps might send you a notification when it's your turn in a game or when you receive a game challenge from another player.

Games on Facebook

The most common type of app on Facebook is a game, so much so that Facebook has a designated Games section in the left sidebar of the Home page. Games can range from matching games such as Bejeweled Stars to board games such as chess to role-playing games such as Modern Princess Life.

Playing instant games on Facebook

Head to the left sidebar on your Home page and click Games. Then click Play Games in the left sidebar of the Games home page to see a list of instant games, as shown in Figure 15-1. If you want to narrow your search, scroll down the left sidebar to find a list of game categories. *Instant games* are games you can play by yourself or in Facebook Messenger with your friends on your computer or phone. Click the grey Play button below a game to launch the game on your screen. You can also click the right arrow to post a story about the game in News Feed and invite friends to play. If you invite a friend to play an instant game, the game will appear in a Facebook Messenger thread between the two of you.

If you click to play a game on Facebook, you see the Game Permissions window, shown in Figure 15-2. This window is the moment of truth in terms of deciding whether you want to share your information with the game. Be sure to read the fine print that goes over exactly what information the app will have access to. Often the info is just your public profile, as mentioned earlier. Sometimes it includes other pieces of information, such as your email address, birthday, or friends list. You generally can't pick and choose what info you share here, but you may be able to do so after you start playing by adjusting your settings, which we describe in the "Adjusting your app permissions" section, later in this chapter.

FIGURE 15-1:
So many games to choose from.

FIGURE 15-2:
Consent to sharing your info with the game.

REMEMBER

You can't avoid sharing your public profile with a game. If you don't want to share your basic information with a game, you can't play it.

If you want more information about the game's developer, click the icon that resembles a card with an arrow pointing toward it (in the lower right in Figure 15-2) to review that developer or company's privacy policy, which all apps and games are required to include on the Game Permissions window. When you have reviewed this information and are comfortable with what you're about to share with the game and how that information can be used, click the big blue Play Now button to load and launch the game (the button may instead say Continue with Facebook). When the game loads, the rest of the screen darkens so that you'll be able to only see the game, but rest assured that you haven't left Facebook and that you can return to the Games page at any time by clicking the X in the upper-right corner of the game.

Playing web games on Facebook

If you find you're really into playing games on Facebook, the Web Games section can be a great place to find out about new games. To get there, click Web Games in the left sidebar of the Games home page. You can search for games in the upper-right corner of the page or use the top bar to sort by Top Charts (most popular games), Casual (puzzles and card games), Battle, Casino, and Real Money games. Gambling laws differ by country and real money games may not be available where you live. If they are, make sure you're comfortable gambling with your hard-earned cash. Web games can be played on Facebook and your mobile phone but, unlike instant games, can't be played in Messenger.

If you want to proceed with a game, click its title. For some games, you'll be taken directly to the permissions page. For other games, you see a pop-up window with more information and a Play Now button. Use the arrows in the upper portion of the window to advance through slides containing screen shots and information about how to play the game. You can use the buttons at the top of the window to like or share the game, or to visit the game's Page on Facebook. If you decide you don't want to play the game after all, click the X in the upper-right corner to close the window and return to the Web Games page. If you want to continue, click the big blue Play Now button.

You'll eventually see the Game Permissions window, whether you get there from clicking the game's title or clicking the game's Play Now button. (In the rare case when the game doesn't require information beyond your public profile, the game will start immediately.) The web Game Permissions window has some additional settings that the instant Games Permissions window does not. The setting we're interested in now is Edit This link. Clicking this link displays a window that itemizes the app's requested permissions. You see which information is required (often your name and email address) and which information you can choose to share or not share. To unshare a piece of information, such as your friends list, simply click the slider button to off. You can then click the big blue Continue as <Your Name> button to begin playing the game.

Inviting and notifying

Just like you can invite friends to events, you can also invite friends to play games with you. After your friends are playing the same game as you or using the same application, you can send them requests for specific actions, as shown in Figure 15-3. Game invites and requests appear in your friend's notifications on the Home page or perhaps in the game itself if your friend is already playing that game.

Posting

Most games won't post to Facebook on your behalf without your express permission in the moment. However, as you play games, you may be prompted to post things to your timeline. Figure 15-4 shows what one such prompt might look like. In this case, a game is suggesting that you invite your friends to play. If you want to share your game with friends, click the Share Now button and feel free to add your own comments to the text field.

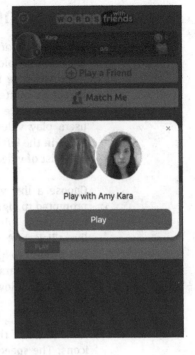

FIGURE 15-3:
Sending app requests.

If you'd rather not post something like this to your timeline, either ignore the pop-up until it disappears in a few moments or click the X in the top-right corner to close it.

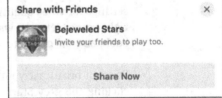

FIGURE 15-4:
Posting to your timeline.

TIP

Instant games have a fun feature where you can capture a clip of your gameplay and post it to your News Feed. To do so, click the camera icon to the right of the game screen and wait while your game capture is created. The sample News Feed story appears and you can click Share Now or click the X to remove the clip.

WARNING

Some games display ads in the middle of game play. To return to your game, you need to search for and click the ad's X, as shown in Figure 15-5.

Watching game videos

The evolution of gaming now includes watching other people as they play games through video websites such as YouTube or Twitch. Facebook wanted in on the action, so they created a place where you can watch other users play video games live. Click Gaming Video in the left sidebar on the Home page to see a list of videos that are currently airing.

Choose a live video to watch. You may be prompted to sign up for notifications so that you'll know when this particular user goes live. Click the X in the upper-right corner of this prompt to ignore it. If you want to follow a particular person, you can always click the large blue Follow button on the right side of the screen.

The bottom of the screen has a few different icons. The speech bubble with a slash turns off comments and reactions from fellow viewers as they happen in real-time on the right side of the video screen. The film slate creates a clip that you can save to your computer or post. The gear icon adjusts settings for video quality and closed captions. The square with a filled-in bottom-right corner lets you watch the video while you continue to other parts of Facebook, and the opposite arrows icon switches the video to fullscreen.

FIGURE 15-5:
Click the X to get rid of pesky ads.

You can broadcast your own video by returning to the Gaming Video page and clicking the grey Start Streaming button on the left sidebar. This brings you to Facebook's Creator Studio. We won't go into details here because becoming a member of Creator Studio is above and beyond how most people use Facebook. But we do want you to know it exists if creating a streaming channel for a Page or for gaming is something that you're interested in pursuing.

TIP

Recently, Facebook added the capability to host virtual tournaments among gaming groups of friends. Click the Create Tournament Event button to start (it's below the Start Streaming button in the left sidebar). You must agree to the Facebook's Tournament Organizer terms before you can create your event. (See Chapter 13 for the lowdown on event creation.)

Viewing your gaming activity

Head to the Gaming Activity section from the left sidebar of the games home page if you want a singular view of your gaming activity and achievements and a list of suggested games based on similar games you've played.

Keeping your games close

The left sidebar on your Home page is where you go to get to different parts of Facebook: your groups, News Feed, events, and so on. Links to the games you play most often appear in the Shortcuts section of the sidebar.

You can pin the games you play most often to the top of the Shortcuts section for easy access. To do so, click the Edit link to the right of the Shortcuts section. The Edit Your Shortcuts dialog shown in Figure 15-6 appears. Search for the game you want to pin using the search bar up top or scroll down the list.

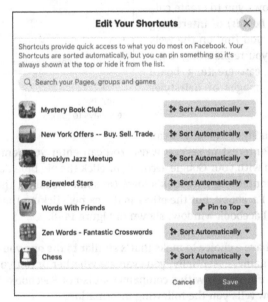

Edit Your Shortcuts

Shortcuts provide quick access to what you do most on Facebook. Your Shortcuts are sorted automatically, but you can pin something so it's always shown at the top or hide it from the list.

Q Search your Pages, groups and games

Mystery Book Club	↕ Sort Automatically ▾
New York Offers -- Buy. Sell. Trade.	↕ Sort Automatically ▾
Brooklyn Jazz Meetup	↕ Sort Automatically ▾
Bejeweled Stars	↕ Sort Automatically ▾
Words With Friends	📌 Pin to Top ▾
Zen Words - Fantastic Crosswords	↕ Sort Automatically ▾
Chess	↕ Sort Automatically ▾

Cancel   Save

FIGURE 15-6:
Edit your
Shortcuts here.

A button with a drop-down menu appears to the right of every listed Page, group, or game. Facebook defaults to sorting these shortcuts automatically, but you can click the menu and choose Pin to Top, so you can always find what you're looking for, or Hide, if you're in the mood to declutter your sidebar. Hiding a shortcut doesn't remove it from your account — you'll just need to use the search bar on the Home page to find it again or head to its respective category in the left sidebar.

Using Facebook Outside Facebook

Believe it or not, there are websites out there that aren't Facebook. Websites like Yelp can help you find a place to eat. Perhaps you want to watch short videos on TikTok. Or you might create pin board collections on Pinterest. Many of the most popular websites allow you to use your Facebook credentials to speed up the process of getting started on a new site, as well as to share activity more widely with your friends on Facebook.

Figure 15-7 shows the landing page for www.pinterest.com. Pinterest is a website that allows you to create collections (called *boards*) of interesting links or images you find on the web. For example, if you're planning to get a haircut, you can create a board where you pin images of hairstyles you like.

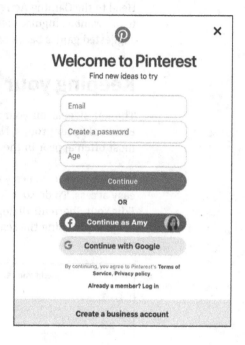

FIGURE 15-7:
A few ways to join.

As you can see in Figure 15-7, you can get started on Pinterest in several ways. You can enter your email and create a password, log in with your Google account, or click the big blue Continue as <*Your Name*> button containing Facebook's logo. On some websites, this button might say Log In with Facebook, but the effect is the same; clicking the button displays the Log In with Facebook window, shown in Figure 15-8.

From here, you have a choice to make that's similar to the one you make when you decide to play a game. Essentially, you evaluate whether or not you want to share your Facebook information with a company that is not Facebook. The Log In with Facebook window tells you the following information:

>> **Your name:** *Your name* is your first name as you use it on Facebook. You see it in the blue Continue as <*Your Name*> button.

>> **The name of the website you're about to log into:** In this example, Pinterest is the website you're about to log into.

FIGURE 15-8:
Log In with
Facebook.

>> **The information that website will receive:** Click the Edit This link to change the information you allow the website to access. At a minimum, you're required to share your public profile.

>> **Information about permission to act:** In other words, the window tells you whether clicking Continue will allow the website to post information back to Facebook on your behalf. In Pinterest's case, you see a closed lock icon and This Doesn't Let the App Post to Facebook, so rest assured that none of your Facebook friends will know that you're filling a pinboard with photos of celebrity haircuts — unless you want them to.

If you're okay sharing this information, click Continue as <*Your Name*> to start using the website. Some websites may ask you to enter your email address after this step so that they can contact you directly instead of through Facebook.

We chose Pinterest because it's a good example of how logging in with Facebook can be convenient. At a minimum, signing up through Facebook saves you the time it takes to create a new account. When you first join, most websites ask you to enter your name and birthday, create a password, and add a profile picture. Signing up through Facebook takes care of that for you (in fact, your Facebook profile picture appears on the site you're using).

SHARING ON FACEBOOK WHEN OFF FACEBOOK

One way to use Facebook when you're using other websites is to share articles, links, images, videos, and websites back to Facebook as a post. Most major news sites or blogs include links to share articles, and sharing on Facebook is usually one of the options. When you click to share content back to Facebook, a window opens that looks like the share box you know and love. You can write a post from there, and when you click the blue Post to Facebook button, the post will be shared with your friends on your timeline and via News Feed or as a story. (Check out Chapter 4 for more on stories.)

Clicking a Share button when not on Facebook works a bit differently than clicking to log in with Facebook or continue with Facebook. When you click a Share button, you're not giving the website access to your information. The same is true if you see a Like button or Send button with a Facebook logo. Those buttons do help you share content back to Facebook, but it's not the same thing as linking your account to a website, an app, or a game.

Perhaps more importantly, sites such as Pinterest have a social element: You likely care about the things your friends have pinned, reviewed, watched, and so on. By logging in with Facebook and giving access to information such as your friends list, Pinterest can quickly help you find people you want to follow or invite to join Pinterest by matching your friends list to their own list of users.

You can also share your Pinterest boards on Facebook, which means even if your Facebook friends don't use Pinterest themselves, they'll be able to see when you've created a new collection you want them to know about.

Mobile Apps and Facebook

If you have an iPhone, an Android, or another smartphone that uses apps, you can connect your Facebook account with mobile apps. The process is similar to the way you connect your account with websites. In fact, because many websites, such as Pinterest, also have mobile apps, you may wind up signing up with Facebook on the web, only to use an app on your phone or vice versa. Similarly, many games you play on Facebook have mobile app counterparts, so you can continue a game on your phone whenever you have to step away from your computer.

REMEMBER

Figure 15-9 shows an example of the mobile Log In with Facebook screen. Like the Log In with Facebook screens you would see when using Facebook on the web, the moment you tap Continue as <Your Name> on this screen on your mobile device is the moment you share your information with the app and link your Facebook account to the app.

And like the permission windows you see on your computer screen, the screen on your mobile device displays the following information:

>> The app's name

>> The information the app will receive

>> Information about whether the app can post to Facebook

When you use Facebook to log into an app on your phone, you reap the same benefits as when you use it to log into a website:

>> Automatic creation of your account and profile without having to enter your details

>> Ability to use your Facebook friends list to find friends in the app

>> Ability to share content from the app back to Facebook quickly and easily when you want to

FIGURE 15-9:
Logging in to an app with Facebook from your smartphone.

Managing Your Games, Websites, and Apps

There are so many apps and games out there that we couldn't even begin to write "Apps and Games For Dummies." You'll need to learn about the games you choose to use by playing them, and the websites you log into by using them. At the same time, because you're using Facebook with these games, websites, and apps, you still control your Facebook information and how it gets used.

For the purposes of discussing how you can change settings related to games, websites, and mobile apps, we refer to all of these incarnations as *apps*. You can view your current settings for any app by following these steps:

1. **Click the down arrow icon (Account menu) in the top bar.**

2. **Click Settings & Privacy from the menu that opens.**

3. **Choose Settings from the drop-down menu.**

4. **In the left sidebar, click Apps and Websites.**

   The top portion of the page displays the web apps and games you've used, websites you've connected to your account, and mobile apps you've logged into with your account. However, this list likely isn't complete because Facebook separates web and instant games. Click Instant Games in the left sidebar to see all the instant games you've used. These instructions apply to both web games and instant games.

5. **Locate the app or website you want to review and click the View and Edit link.**

   The window shown in Figure 15-10 appears, which is an overview of that app or website's settings and the information you're sharing with it.

FIGURE 15-10:
Edit the app's permissions here.

Adjusting your app permissions

At this point, you're viewing the permissions window for the app you clicked. The first section in the app's permissions window is Understanding the Info You're Sharing with the App. This section contains a list of the information that you shared with the app when you started using it. Each piece of information the app requested is listed, and a blue slider signifies that it's currently being shared. Below the information type, you can see specifically what is shared. For example, you might see that you shared your personal email address versus your work email with this app.

You can click the slider to remove any piece of information from the list except your public profile (which, remember, includes your name, profile picture, age bracket, gender, and other information you share publicly).

Making additional app settings

The Additional App Settings section is about whether or not the app can send you notifications. If you want to turn off notifications from this app, simply click the drop-down menu and choose No. There's also a setting for who can see you use this app. The app visibility is controlled by the same privacy menu that you use to choose who can see your timeline posts. Click the drop-down menu to choose whether this information is visible publicly, to just friends, or to a custom set of people.

TIP

Many apps automatically set their visibility to Only Me. So if your friend is also playing Words with Friends but she can't locate your account in the game, try changing this setting to Friends.

Learning more

The Learn More section provides links to the company's privacy policy and your user ID number, which is different for every app. You may need your user ID number if you ever contact the developers of the app for assistance with an issue.

REMEMBER

If you change anything in the app permissions window, click the blue Save button at the bottom of the window to make sure the new settings stick.

Removing apps

The simplest way to adjust how an app interacts with your information is simply to revoke its ability to interact with your information. This option makes sense only if you are 100 percent finished with an app — that is, you don't plan to use it again in the future.

To remove an app, go to the App Settings page (as described in the section "Managing Your Games, Websites, and Apps") and select the box to the right of the app's name. You can do this with as many apps as you want. After everything is checked, click the blue Remove button at the top of the page.

A window pops up explaining that if you proceed, the app will be removed from your Facebook account but may retain some of the data you originally shared with it. (The app just won't get any new non-public data going forward.) If you want to fully purge any past app activity from your account, select the Delete Posts, Videos, or Events These Apps and Websites Posted on Your Timeline option. You can also select the option to let the app developers know that you're removing their app. Clicking the Remove button completes the process. You can see a list of apps you have removed in the Removed tab.

When you remove an app, you can no longer use the app, and it won't be able to send any information or posts to your timeline. At the same time, because so many apps exist outside of Facebook, you have to keep in mind that the app developer will still be able to contact you via your email address and may have their own account created for you. If you want to cut off an app completely, you may need to reach out to the developer and request that they delete your account and information from their servers.

Adjusting your preferences

Below the list of apps you use is a section for preferences. The first, Apps, Websites, and Games, has to do with your ability to interact with apps on and off Facebook. By default, this setting is turned on. If you click the Edit button, you can turn this setting off, but it will remove Platform from your account and you won't be able to interact with apps, websites, and games using Facebook. This nuclear option is for when you are sick and tired of receiving your umpteenth invite from your kind but oblivious aunt to play a game that doesn't interest you.

While you can always undo this option and turn Platform back on, we recommend avoiding the back and forth by managing your notifications and invites with the next setting: Game and App Notifications. This setting controls whether or not you receive those invites to play games or join an app. If you find that your notifications are overrun with game requests and your sanity is hanging by a thread, click Edit and then Turn Off. Problem solved!

Controlling what you see from friends

News Feed can be a great way to discover what apps your friends use, but it can also be inundated with app stories that block out the interesting content unrelated to those games and apps. Here are a few tips to keep your News Feed (and the rest of Facebook) from being cluttered by apps:

» **Hide the app from News Feed.** If your News Feed is inundated with posts from apps, click the three dots icon at the top right of the post. A menu of options appears, including Hide All from *<app>*. When you click Hide, the post disappears from News Feed and is replaced by text confirming that it has been hidden.

» **Block an app.** If you find an app offensive or it keeps sending you invites or requests, you can block it. From the Settings page, navigate to the Blocking tab using the left sidebar and enter the app's name in the text field in the Block Apps section. The app will no longer be able to contact you via Facebook or see any of your non-public info.

» **Block a friend's invites.** Sometimes just one person is the problem. The person may be sending you invites or requests from multiple apps, and it's driving you nuts. Navigate to the Blocking section of the Settings page and enter your friend's name in the text field in the Block App Invites section. Any invites or requests the person sends you will automatically be ignored and won't generate any notifications on your Home page or in your email. You'll still be friends with the person, and you'll still see posts from him such as status updates and photos; you just won't get the app stuff anymore.

Reporting offensive apps

If you are using an app and think it is offensive, using your information inappropriately, spammy, experiencing bugs, or not working properly, you can send a report to either Facebook or the app developer. Launch the app and locate the three dots icon at the top-right corner of the screen. Click the Report to Facebook option. A window opens with possible reasons for reporting the app. Depending on what you choose, you'll be prompted to upload screenshots and send the report so that Facebook can review it and determine if any of its terms or policies are being violated.

5

The Part of Tens

IN THIS PART . . .

Some new — and tried and true — ways to get the most out of Facebook

Answers to common questions

Ways to participate in politics on Facebook

Chapter 16

Ten Ways to Make the Most of Your Facebook Content

We spend a lot of time in this book emphasizing how easy things are to do or explaining the simplest way to do things. Want to update your status? Click, type, post. Want to post a picture from your phone? Tap, select, post. Using Facebook is meant to be easy and seamless when you want it to be. This chapter, however, is about going beyond the basics. We suggest ways to customize your content or to take stock of your history on Facebook and bask in the nostalgia. Ultimately, these tips will help your posts stand out in a sea of News Feed stories.

Remembering the Past

Quick, what were you doing a year ago? Not, like, in general. But exactly 365 days ago. Were you having a good day or a bad one? Did you go to the grocery store or catch a movie? If you've been using Facebook for over a year and, like most people, use it every day, you can find out by revisiting your Memories. Head to the left sidebar on your Home page and click Memories to see every post you made "on this day" since you became a Facebook member.

Depending how long you've been on Facebook, it could be from several years ago or from just last year. Your memory might be a photo, a status post, or a timeline post someone left for you. Sometimes it might be a bit boring, but at the same time it's fascinating to see how things have or haven't changed. And sometimes it brings up something so delightful you can't believe you've forgotten it — a funny photo you took or the sweet words of a friend. You can post these discoveries again by clicking the word Share below the memory. The share box appears so you can add context to the memory or details such as location before posting it to your timeline. Some examples of memories you might post include reflecting on the birth of your grandchild now that she's a toddler or laughing all over again at a clever joke your friend made. Go ahead and check the Memories page whenever you're in need of some nostalgia. Facebook will also occasionally post a memory to the top of your News Feed (it won't be shared with your friends unless you choose to share it with them).

Scrapbooking Baby Photos (Mobile Only)

If you're the parent of young kids and choose to share photos of them on Facebook, you might run into a problem where you and your spouse or partner are constantly adding one-off photos of the cuteness but then having trouble locating those photos. Facebook created a special way to compile photos of your kids called Scrapbook. Scrapbook lets you create tags for your children without creating a profile for them. You can then tag your kids in photos. Only you and your partner can tag your children, so you don't have to worry about them being tagged in photos without your permission.

To create a scrapbook, navigate to the About section of your timeline and click to add a family member in the Family and Relationships section. When you add your child, you can turn Add Scrapbook on or leave it off. When you click Save, you'll be prompted to get started on your scrapbook and tag photos you've already added to Facebook. After you've created your scrapbook, you can keep adding to it by continuing to tag your children in posts. Click your Photos, then your Albums, and then your child's Scrapbook to view all the photos in one place.

Framing Your Profile Picture

If you spend enough time on Facebook, you'll eventually see your friends update their profile photos with specialty frames. Often, these frames highlight a cause or holiday. But don't let that stop you from adding frames whenever your heart desires. Head to your profile and click the Edit Profile button. Then click the Edit link next to your profile picture. Click the Add Frame button to browse the multitude of available frames. You may need to reupload your profile photo for a preview of the frame. See Figure 16-1 for an idea of how frames appear. Once you've landed on the perfect frame, decide how long you'd like it to stick around (the most popular options are 1 week or 1 month) and click the Use as Profile Picture button to make the switch. This should trigger a News Feed story that your friends will see.

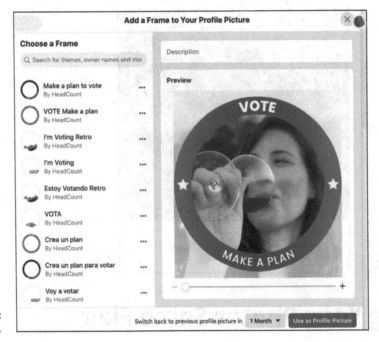

FIGURE 16-1:
Frame-up job.

Adding Dimension

If you want to take your photos to the next level, make them three dimensional. Head to the share box to post a photo as you normally would. Choose a photo, click Make 3D in the upper-left corner of the photo, and then post as normal. The photo appears to move and gain a third dimension as you scroll past it in News Feed.

Your friends can see your 3D photos regardless of where they are viewing Facebook — on mobile, desktop, or tablet — but you can post these photos only from the mobile Facebook application.

If you think 3D photos are cool, just wait until you post your photos and videos in full 360 degrees. Start with a panoramic photo or video and upload as usual with the share box. Your posted photo or video contains instructions to drag the mouse to look around (see Figure 16-2). Clicking and dragging allows you to experience the entire photo as if you were standing where the photo was taken and slowly turning around.

FIGURE 16-2:
Living in
360 degrees.

Giving Your Photos Some Flair

When you post a photo from the share box, you can add creative touches by clicking the pencil icon (edit). The editing options are just that — options. They aren't required, and no one will mind if you just post your photo as a photo. At the same time, these options can just make your photo a little more exceptional.

Options include adding filters, text, and stickers, as well as cropping and rotating your photo. Keep in mind that the full suite of options is only available on the mobile version of Facebook. Computer options are mostly limited to crop and rotate. For specifics on how to add these pieces of flair to your photos, check out Chapter 11.

Reviewing the Last Year (or Years)

Every winter Facebook offers the ability to look back on everything that's happened to you in the last year. Often played as a slideshow, Year in Review videos highlight the most important parts of your year — the milestones are compiled into a digital album, so no trip to the scrapbook store is required.

Facebook periodically makes other sorts of review videos for various occasions. From time to time on its birthday (February 4), Facebook offers the ability to look back through everything you've done since you joined Facebook. Watch the top of your News Feed for a Facebook promotion for this type of video. You aren't required to post these videos to your own timeline, but they can be fun.

Making Your Status Stand Out

By default, your status posts are displayed as black text in a white box. There's nothing wrong with that. But if you want to draw more attention to your post, you can easily add a background to it. When you add a background, your post is displayed centered and bolded with the background you choose behind it.

To add a background to your post, click the multicolored box in the lower-left side of the share box to open a list of background options. Click one of the thumbnail previews to choose a background. Some backgrounds have designs, such as raindrops or birthday cake slices, and others are simply a color. Check out Figure 16-3 for a sample background post. When you're happy with both your background and your text, click Post.

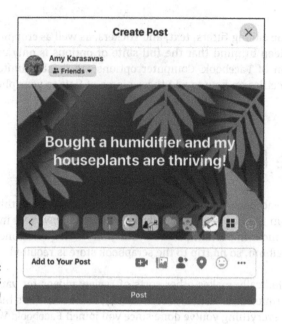

FIGURE 16-3:
Make your posts
pop with
backgrounds.

Tagging It All

In addition to tagging friends in photos and posts, you can tag almost any *thing* in your posts. You can tag celebrities, television shows, movies, bands, companies. Honestly, almost anything. To start tagging, simply type the @ symbol and begin typing the name of the person, place, or thing you want to tag. As you type, Facebook will autocomplete with suggestions. When you see the one you want, click it. So if you're excited about the new *James Bond*, you can tag it, and when friends mouse over the tag, they will see a preview of its Page. They can then click the tag to view the *James Bond* Page and check out all the cool movie posters. You don't need to have previously liked (or followed) a Page to tag it; just type @ and go.

Using Stickers or GIFs in Your Messages

In conversations, there's always much more going on than what's said. There are gestures, expressions, emphasis. No matter how fast we get at communicating via text, something always gets left behind. Enter stickers and GIFs, which are visual ways to represent sentiments. Using them can function as a punctuation to the messages you send to friends. Click the sticker or GIF icon at the bottom of a chat window to browse the hundreds upon hundreds of options Facebook offers. Stickers and GIFs can represent emotions, activities, people, places . . . anything really.

Friend-a-versaries

One of the Memories you may occasionally see is the memory of the day you first became friends with someone important in your life. Congratulations, you can celebrate your friend-a-versary on Facebook! The easiest way to celebrate is to share the post about your friendship's ripe old age and include a little note about why that friend is important to you, or what you think about the fact that you two have been friends for so many years.

It's important to note that Facebook marks the anniversary of the day you became friends on Facebook. If you were friends long before that, your friend-a-versary might seem inaccurate. Still, it never hurts to tell a friend how happy you are to be friends.

Chapter **17**

Ten Ways to Be Politically Active on Facebook

I n today's charged political climate, it's inevitable that you'll come across political content on Facebook, whether from strangers or friends. You don't need to engage with this content; you can simply ignore it or adjust your News Feed settings to see less of it. However, if you want to exercise your political muscles, Facebook offers many ways to get involved, from understanding your voting rights to sending your local representative a message. In this chapter, we go over some ways you can be politically active on Facebook and provide recommendations for avoiding conflict with those who hold different views.

Familiarize Yourself with the Voting Information Center

From the left sidebar of the Home page, navigate to the Voting Information Center to see the latest news about federal, state, and local elections. Figure 17-1 shows you what this page looked like on Election Day. On the right side of the page are election posts with tailored information for you, such as the dates for any special elections in your district. If there's no local news, you may see historical facts instead. (At the time of writing, this page displayed information about the 46th President, the current Senate and House seat numbers, and information about voting from abroad.)

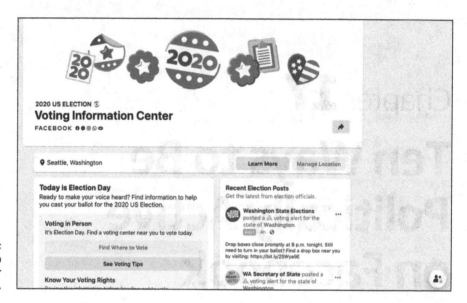

FIGURE 17-1:
Your one-stop shop for voting info.

A few weeks before a state or federal voter registration deadline, you'll see a prompt in your News Feed to check your voter registration status. If you're not yet registered to vote, you can click the prompt and sign up through your state's voter website without having to leave Facebook.

After you're registered, you'll see additional prompts about the upcoming election, including information about your voting rights. For example, you might see a reminder that as long as you're already in line when the polling location closes, you're allowed to cast your vote. You can also identify your polling location based on your address, learn how to request an absentee ballot if you're unable to vote in person, and follow federal voting results in real-time.

Share Your Voting Status

On Election Day, you'll see a News Feed prompt to create a post and let your friends know that you are voting or have voted (see Figure 17-2). This popular feature has become a badge of honor for people to celebrate their participation in the democratic process. It's essentially the digital equivalent of one of those I Voted stickers. It's up to you whether or not you want to create a post to share this information.

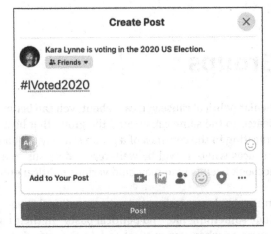

FIGURE 17-2:
I voted!

Spend Time in the Town Hall

From the left sidebar of your Home page, navigate to Town Hall. Town Hall provides a myriad of ways to engage with your elected representatives. From here, click the blue Enter Address button. A pop-up window appears and prompts you to enter your address to determine who represents your district. This information will not be displayed to your friends. After filling out your address, click the blue Submit button to see a list of Facebook Pages for government bodies such as your local Parks and Recreation department and individual representatives such as your state Senators.

The list is divided by Government Services, Local Representatives, State Representatives, and Federal Representatives. Click the blue Follow button to follow their Page and begin receiving updates in your News Feed. You can also click the grey Contact button for a pop-up screen with their work address, phone number, and email address. You may see up to three buttons at the bottom of the pop-up with options to email, call, or send them a message via Facebook messenger.

On the right side of the page are three sections that we find particularly useful. The first is the ability to turn on your constituent badge. To do this, navigate to the Constituent Badge box and choose On from the drop-down menu. Now, if you comment on a post from one of your elected representatives, the badge will identify you as a constituent and signal that he or she should address your questions or concerns. The next box is the Voting Reminders box. Choose On from the drop-down menu to receive voting reminder notifications about upcoming elections in your area. Then head to the Town Hall Updates box and choose On from the drop-down menu to receive updates from your government representatives in your News Feed.

Join Relevant Groups

If there's a particular political cause you care about, you can bet multiple Facebook groups are dedicated to the same cause. Find the group that best fits and request to join. Communicating in the confines of a group is a way to channel your political energy into spaces where it will be well-received versus to a wider audience such as your Facebook friends who may hold various political opinions.

Some groups on Facebook are designated as non-political spaces. Many of these groups will include something like "no political posts" in their list of group rules that you must agree to when you request to join the group. Be aware of which groups you belong to that invite political discourse and which do not.

Fundraise for a Cause

We go over the ins and outs of fundraisers in Chapter 12, but one of the ways you can have an effect is to set up a fundraiser for a non-profit you care about. Any fundraisers you create are prominently displayed to your friends in News Feed, and you can add personal details to explain why this particular cause is near and dear to your heart. Many people set up fundraisers around their birthday and request donations in lieu of gifts or cards.

Organize a Digital Grassroots Campaign

Grassroots activism is responsible for some huge moments in our history. Some great (and not so great) movements began their life on Facebook. Let's say you want to campaign for an issue in the hopes of getting it on the ballot for your state election. You can create a Page to share information and updates, a group to organize leadership, and an online event for people who want to learn more about the issue. Facebook provides multiple channels to serve as a megaphone and attract others to your cause.

Fact-Check Your Sources

Facebook is attempting to curb the spread of misinformation through the site by employing fact-checkers who flag fallacious content. If you attempt to post a link or video that has been flagged, you'll see a pop-up window warning you that the content contains misinformation. You can choose to post it anyway, but it will appear lower in News Feed, greatly reducing the number of people who see it. In extreme cases, Facebook may even remove content from the site if it incites violence or is deemed dangerous to national security.

You might also see a warning on a friend's post that it has been found to contain false information. In this case, Facebook is already aware that this content is not reliable. If you come across a post that spreads misinformation and does not have a warning, you can report it by clicking the three dots icon in the top-right corner of the post, choosing Find Support or Report Post from the drop-down menu, and then selecting False Information as your reason for reporting.

One way to defend against the spread of misinformation is to take responsibility and fact-check your sources before posting something to Facebook. Before you post, do some research on Google, check snopes.com (a popular debunking website), and consider where you saw the information. Was it from a reputable source such as the *New York Times* or a less reputable source such as a personal blog? If you're not sure, you can always err on the side of not posting.

Beware of Click-Bait

You may find your News Feed peppered with stories that seem too unbelievable to be true. Often, that's exactly the case. Much like our advice in the preceding section to fact-check your sources, we also advise that you don't give credence to every article or video you come across on Facebook. *Click-bait* is content designed to hook you immediately so that you feel compelled to click the link. This sneaky marketing trick increases a website's visitor numbers. If something seems interesting to you, check it out. Just be wary of things designed to infuriate or shock you. Odds are high that they'll do just that and cheapen your Facebook experience in the process.

Don't Feed the Trolls

Perhaps you've heard of the troll phenomenon. A *troll* is a person whose mission is to fan the flames of tense discourse and spiral every online encounter into a screaming match. Add politics to the mix and it's easy to lose your cool. Trolls exist on Facebook as well as every other corner of the Internet. But there's a secret to dealing with them: Do not engage. It sounds simplistic, but trolls are bullies who feed off of attention. Don't respond to their baiting comments. Of course this advice is much more difficult to follow if said troll is a friend or family member, but the same principle applies. Trust us when we say that online arguments never ever end well.

Mute What Drives You Crazy

At the end of the day, Facebook should be a place where you communicate with friends, nurture your interests, and make your life easier. Not every interaction needs to be sunshine and rainbows, but if you find that using Facebook is starting to feel more irritating than joyous, it's time to mute the people who are driving you crazy. One of the best ways to accomplish this Facebook feng shui is to adjust your News Feed settings and mute the person (check out Chapter 4 for details). You can mute someone for 30 days or indefinitely. You will remain Facebook friends, but you will no longer see their content in your News Feed.

Chapter **18**

Ten Frequently Asked Questions

H aving worked for Facebook and on this book for many years, we know a lot about the specific complications, confusions, and pain-points people come across while using Facebook. At dinner parties, group functions, family events, or even walking across the street wearing a Facebook hoodie, someone always has a suggestion or a question about how to use the site.

This behavior is understandable. Facebook is a complex and powerful tool with a ton of social nuances, many of which have yet to be standardized. Facebook has a lot of features — and those features change frequently. Each year, Facebook modifies parts of the site, redesigns how certain pages look and feel, and adds features. To keep up on what's happening with Facebook, like the official Facebook Page, found at www.facebook.com/facebook, and you'll get updates straight from the horse's mouth.

What follows are the questions we hear most often from friends and family (and the occasional message from a stranger who really needs help), often with strain in their voices or pain in their eyes. By highlighting these questions, we hope to save you the stress of encountering these issues and wondering whether you're the only one who just doesn't get it.

Do People Know When I Look at Their Timelines?

No. No. No. When people see stories about their friends pop up on their Home page, they sometimes get a little anxious, thinking that this means Facebook is tracking everything everyone does and publishing it to everyone else. That's not true.

Consider two types of actions on Facebook: creating content and viewing it. Creating content means you've intentionally added something to Facebook for others to look at or read, such as uploading a photo or a video, commenting on or liking something, or posting a status. These types of actions are all publishable posts — that is, stories about them may end up on your timeline or in your friends' News Feeds — although you have direct control over who gets to see these posts.

The other type of action on Facebook is viewing content, such as flipping through photos, watching a video, clicking a link your friend has liked, or checking out someone's timeline. Unless someone is looking over your shoulder as you browse, this type of action is strictly private. No one is directly notified about your action, and no trace of the fact that you took that action is left on your timeline or in your friends' News Feeds. So now you can check people out to your heart's content.

I Friended Too Many People and Now I Don't Like Sharing Stuff — What Can I Do?

Having a big friends list is a sad reason to not be sharing with people you *want* to share with. You can fix this problem by using privacy settings, which are covered in detail in Chapter 6. Here's an overview for how to change who can see a post you are making:

1. **When you've completed your post, click the privacy menu by your name in the share box.**

   The menu usually says Public or Friends by default.

2. **In the menu that appears, choose Friends Except.**

   A window appears for choosing people from your friends list.

3. **Click the faces of the friends whom you *don't* want to see your post.**

   To find specific people, scroll up and down or type their names in the search box at the top. You can select as many people as you want.

4. **When you've finished choosing people, click Save Changes.**

   The privacy setting you chose in Step 2 is now your new default for when you share posts with people.

Another thing you can do is start to remove people from your friends list if you don't think it will cause social awkwardness in real life. Most people, especially old work acquaintances, don't notice when they've been removed from someone's friends list.

Facebook Looks Different — Can I Change It Back?

Inevitably, Facebook is going to change the way it looks. You're going to log in one day, and things will look different — the features you were used to seeing on the left will now be on the right, or moved who knows where, or gone completely. It's confusing. Facebook changes the look and feel of the Home page or the timeline about once a year. And trust us, the day that you log in and find that this has happened, you're going to hate it.

Unfortunately, no matter what you do, no matter how much you hate it, Facebook rarely goes back on a redesign and you can't change it back. The best thing to do is to try to figure out the new site. Check out the Help Center (open the Account menu in the top bar, click Help and Support, and then click Help Center) or Facebook's Page (www.facebook.com/facebook) to read about the layout changes and how the site works. And then try to use Facebook a few minutes a day until you get used to it. Over time, it won't seem so bad. You'll look at a photo of the old Facebook and think how ugly it looks by comparison.

So, short answer: No, you can't change it back. But we have complete confidence in your ability to adapt to the new Facebook.

I Have a Problem with My Account — Can You Help Me?

We wish we could help. Unfortunately, we are but users like you, and that means although we can help diagnose the issue, we can't usually treat it. Sometimes the problems are Facebook's fault, and sometimes they are user errors, but either way, we don't have the tools required to fix them. Most account problems can be resolved only by Facebook employees with special access to the specific tool required to fix an account. Here are a few of the account questions we've received recently, and the answers given:

>> **I can't remember my password. Can you reset it for me?** The answer: No can do. Click the Forgot Account link on the login page to start the reset process, which entails Facebook sending a password reset code to your email or Google account.

>> **My account was deactivated because it said I was sending too many messages. Why? Can you fix it?** Answer: We recently had this happen to two friends, one who was using his account to promote his music career, and one who was distributing his poetry to many, many friends through messages. The result was Facebook spam detection at work. When an account starts sending a lot of messages in quick succession, especially when those messages contain links, it looks a lot like spam to the system.

In most cases, the person is warned first, but if the behavior continues, his account is disabled. The only way to have this action reversed is to write in to Facebook's Help Team and request reactivation. To write to Facebook, go to the Help Center (www.facebook.com/help) and search for a FAQ entitled My Personal Facebook Account Is Disabled. Follow the instructions for contacting Facebook. The process of getting your account reactivated can sometimes take several days.

What Do I Do with Friend Requests I Don't Want to Accept?

This is a tough question. A social convention doesn't yet exist for this situation, so the answer to this question is personal. Just know that there are several actions you can take:

>> **Click Delete Request.** Remember, people are never notified if you've rejected their friend request. If you don't want to be the person's friend, you don't have to be.

>> **Leave the request sitting there forever.** Sometimes you just aren't sure if you want to be friends with someone, but you also aren't sure that you don't want to be friends with them. However, if you're the sort of person who has thousands of unread emails in your inbox that you just aren't ready to deal with, this solution might not work for you.

>> **Accept the request and then add the person to a special restricted friends list.** This approach works if you don't want to accept the friend request only because you don't want that person having access to your timeline. You can go into your privacy settings and exclude that friends list from seeing any parts of your timeline that aren't set to public. Then anyone you add to that list will be restricted. In this way, you can accept the friend request without giving access to your timeline.

>> **Click Confirm.** This simple solution works if you don't want to accept the request only because you don't want to read about that person in your News Feed. The first time she shows up in News Feed, hover your cursor over the post and click the three dots icon in the upper-right corner of the post. Choose Unfollow *friend's name* in the menu that opens to prevent any future stories from that person from appearing in your News Feed.

Why Can't I Find My Friend?

Let's assume you're asking this question after exhausting every possibility for finding friends, as described in Chapter 8. And we'll also assume that you're looking for a specific person, not friends in general.

You won't be able to find a friend for the following few reasons:

>> **She hasn't joined Facebook.** Shocking, we know. If you have her email address and you think she'd enjoy Facebook, you can always invite her to join and be your friend.

>> **She goes by a different name on Facebook to protect her privacy.** For example, if her name is Jane Smith, she may list her name as Janie S. Try searching for her by her email address or phone number.

>> **She has a common name.** Facebook Search tries to get you to the right Jane Smith by looking at information such as friends in common and shared hometowns, but sometimes it comes up empty.

>> **She doesn't have much information filled out on her timeline.** If you're looking for a high school classmate but she never entered her high school information, you're going to have a hard time finding her.

>> **She blocked you.** Yes, this one is harsh. We put it on the list only because we've seen it happen before. Someone says to me, "I *know* she's on Facebook. And I *know* she's friends with my friend. But when I go to find her she's not there." While it hurts to be blocked by someone, don't drive yourself crazy looking for reasons why it happened. If she doesn't want to connect with you on Facebook, that's her loss; move on to your other friends and all the things you can share with them.

Will Facebook Start Charging Me to Use the Site?

Another simple answer: No. This rumor is a particularly nasty one that makes the rounds every now and again via people's posts. Several variations exist, but they usually involve asking you to repost the status that Facebook is shutting down or going to start charging or running out of names. Don't fall victim to this ruse. Facebook has long maintained that it will always be free to users. Unless you're advertising something, Facebook will always have space for you for free.

How Do I Convince My Friends to Join?

While the obvious answer to this question is to give them a copy of this brilliant book about how to use Facebook, you can also try a few other things. You can tell her anecdotally the ways in which Facebook has enriched your life. Maybe you are interacting with your kids more, are keeping in touch with friends you thought were lost, or have a place to put your thoughts and photos where your friends might see them. You can let her look over your shoulder as you use the site so that she can see the experience herself — ask her questions about whether there's anyone she'd like to look up. The more information she sees about the people she cares about, the more likely she is to take the next step.

One common complaint from people who haven't joined the site is that they don't have time for yet another online thing. To this concern, one common response is that Facebook is an efficiency tool that often saves a person time compared to using old-school methods of communication. Messaging can often replace email, and events are easier to coordinate over Facebook. Sharing phone numbers is easier. Sending and receiving links is easier. Finding rides to the airport, restaurant recommendations, and who is heading to the park on Saturday are all faster and easier than trying to use email, phone, or other methods of communication.

REMEMBER

Finally, for some people, it's just not their time. No matter what you say, they'll stick their fingers in their ears and sing *la-la-la* until you start talking about sports or the weather or the circus coming to town next week. You can't force them to Facebook; you have to let Facebook come to them. Over the years, we've watched many a nonbeliever eventually cross over and discover the value in Facebook. Patience may be your only weapon for these diehards.

What If I Don't Want Everyone Knowing My Business?

To those who ask that question and don't have time to read Chapter 6, which goes into detail about how to be a private person on Facebook, we will simply try to impart the following message: You can be an extremely private person and still derive nearly all the same value out of Facebook as anyone else. All you have to do is learn how to use the privacy controls and lock down all your information and access to your timeline, ensuring that only those you trust can see your info. From there, you can interact in all the same ways as anyone else without feeling as though your privacy is being compromised.

Note: Besides understanding the privacy settings and taking the initial time to adjust yours until they feel just right, you have to do a little extra work to be private on Facebook and still derive comparable value. You'll likely have to put in extra effort connecting with friends because the more locked-down your information, the harder it is for not-yet-Facebook-friends to find your timeline, and the harder it is for your friends to find, identify, and connect with you. If you're willing to do the work of seeking out your friends and connecting with them, however, your experience should be nearly identical with everyone else's.

Does Facebook Have a Feature That Lets Me Lock Myself Out for a Few Hours?

Short answer: Not really. Long answer: People deal with Facebook overload in several ways. One is *deactivation,* in which you shut down your account temporarily. It means no one will see your timeline or be able to interact with you on Facebook. The benefit of such an action is that you're guaranteed not to get notifications about messages, picture tags, timeline posts, or anything else. The downside is that it will cause a lot of confusion among your friends who suddenly can't message you, tag you, or write on your timeline. If they have your email address, they're likely to bug you anyway to ask why you disappeared from Facebook.

To reactivate at any time, just enter your password (just like signing in). Your account will be completely back to normal. If you're remotely curious about how your social group has evolved without you, you may have trouble truly staying away.

Another option people choose if they are big mobile users is to delete the Facebook app from their phone so that their Facebook experience happens only on a computer. This approach cuts down on mindless scrolling while they're out in the world and bored. Even your intrepid authors have been known to delete the Facebook app from time to time. Similar to deactivation, when things changed and they felt like they wanted it back, they redownloaded it and were off to the races!

Just like many good things in life, the key to keeping them good is moderation. French fries are delicious, but too many give you a tummy ache. Dancing is a blast 'til your feet are covered with blisters. Television is educational and entertaining until it's 3 a.m and you are watching your fifth infomercial. Facebook is no different. It's a brilliant utility when used to make your life easier and your social interactions richer. When you find yourself flipping through two-year-old vacation photos of a friend of a friend of a friend of a friend, it's time to blink a few times, step away from the screen, and go out for ice cream, or dancing, or whatever it is that gives you joy.

Index

Numbers

3D photos/videos, 171, 377–378

A

About page, 46
About section (Pages), 322
About You section, 105
accepting friend requests, 31–32, 202
Account icon (top bar), 39
accounts. *See also* timelines
  deactivating, 126, 396
  deleting, 126
  fixing problems with, 392
  keeping secure, 117, 148
  logging in to websites/apps with, 123, 364–367
  managing account information, 126
  recovering, 124
  transferring media from, 125
  verifying through friends, 123
Active Contacts option (Messenger), 226
Active Status setting (Messenger), 187
Activity Log
  accessing, 106
  for Groups, 251
  privacy and, 150–152
  settings for, 354
Ad Account Contact (Settings page), 122
Ad Center section (Pages), 346

Ad Choices link, 46
Ad Limits setting (Pages), 352
Add Members option (Messenger), 219
Add Messenger to Your Website option (Pages), 350
Add Organizers option (fundraising), 306
Add to Your Post window, 78
administrators
  for Groups, 248–256
    adjusting settings, 252–253
    editing members, 253–255
    interpreting insights, 253
    managing Groups, 250–252
    pinning announcements, 249–250
    reporting offensive groups and posts, 255–256
    scheduling posts, 249
  for Pages
    messaging, 321
    role of, 351
Admins and Moderator option (Groups), 253
Ads Shown Off of Facebook setting (Ad Settings section), 145
Advanced Messaging setting (Pages), 351
advertisements
  creating, 38
  in Marketplace, 288
  in News Feed, 59
  posting as Pages, 336

privacy settings for, 142–145
Profile Information section (Settings page), 119–120
shortcuts for privacy settings, 148
Social Interactions setting, 120
Advertisers (Page role), 351
Advertising link, 46
age limits, 17
Age Restrictions setting (Pages), 348
album covers, 277
albums
  adding photos to, 174
  automatic, 278
  creating, 271–273
  deleting, 278
  editing, 274–275
  viewing, 262–263
alerts, when logging in from unfamiliar places, 117
All Posts option (Groups), 243
Allow Others to Share Your Posts to Their Stories? setting (Profile and Tagging section), 130
Allow Others to Share Your Stories to Their Own Stories? setting (Stories section), 138
Also from Facebook icon, 164
alternative text for photos, 267
Always Show Captions setting (Videos section), 146
Analysts (Page role), 351
angry icon, 62

apps. *See also* Facebook app; Messenger

accessing personal information, 356–358

adjusting permissions of, 369

blocking, 136

controlling in News Feed, 371

Learn More section, 369

managing, 367–371

permitting to take actions, 358

posts by, 59

preferences for, 370

removing, 369–370

reporting, 371

settings for, 369

third-party

blocking, 371

hiding posts about, 371

overview, 14–15

posting on Facebook, 358

sending notifications, 358

Apps and Websites section (Settings page), 119, 140

Apps setting (group), 252

archiving posts, 96

Ask for Recommendations option (Facebook app), 169

attachments, sending, 217

Audience Optimization for Posts setting (Pages), 348

Audience-based Advertising setting (Ad Settings section), 144

Audio Call option (Messenger), 224

Authorized Logins, 123

autocomplete, 33, 40, 65

automated responses, 341–342

Automatic Member Approvals tool (Groups), 251

auto-play videos, 146

avatars, 162, 183

B

Background Color (Facebook app), 169

Badges setting (Group), 252

banning users, 352

bios, 89

birthdays

contact information, 24, 102–103

viewing in Events, 311

Birthdays section, 47

blocking

app invitations, 135

apps, 136

event invitations, 135–136

invitations, 371

messages, 135, 220

Pages, 136

restricted list, 133–134

settings management, 133–136, 222

third-party apps, 371

users, 116, 134–135

Blood Donations section, 105, 162

Boomerang option, 176, 179

Branded Content setting (Pages), 352–353

Business App Store section (Pages), 342

Business Integrations section (Settings page), 142

buying in Marketplace, 286–289

Buy/Sell groups, 248, 297–299

C

call-to-action button, 321, 329

camera, 175–177

Campus icon, 163

canceling events, 316

Captions Display setting (Videos section), 146

captions for videos, 332

care icon, 62

Careers link, 46

Categories setting (Events), 312

Categories Used to Reach You setting (Ad Settings section), 143–144

Change Password section, 122–123

chat list, 221–222

Chats icon (Messenger), 188

check-ins

defined, 56, 169

viewing locations where you've checked in, 106

City filter (friends), 206

click-bait, 387

Climate Science Information Center, 163

Color option (Messenger), 218

Comment Ranking setting (Page settings), 349

Comment Ranking setting (Public Posts section), 132

comments

adding, 63–65

censoring words from, 130

editing, 65

GIFs and, 64

in Groups, 242

liking, 65–66

Pages and, 336–337

photos and, 64

in posts, 54, 166

posts about, 58

removing, 65

replying to, 65–66

stickers, 64

tagging, 64–65

community guidelines, 16–18

Community tab, 321

confirmation email, 25–26

connecting with Pages, 322–324

Constituent Badge box (Town Hall), 385–386

contact information, 102–103, 121–122

Contacts section, 47

Content Distribution setting (Pages), 349

content in News Feed, 53

Continue Watching while You Use Facebook icon, 279

Control Face Recognition shortcut, 146

Conversation Name option (Messenger), 219

conversations in inbox, 225–226

convincing friends to join Facebook, 394–395

Cookies link, 46

Country Restrictions setting (Pages), 348

cover photos
adding, 247, 327
defined, 84
on mobile browser, 193
for Pages, 320–321
posts about, 58
in profile, 85–86

cover videos, 327

COVID-19 Information Center, 163

Create Group option (Messenger), 219

Create icon (top bar), 38

Create Room option (Facebook app), 169

Create Tournament button (Games), 362

Crisis Response (shortcut in sidebar), 44, 163

cropping photos, 172, 267

Crossposting setting (Pages), 354

Custom (privacy setting), 112–113, 283

Custom posts, 81–82

D

Daily Text Limit setting (Facebook Text), 196

Dark Mode setting (Messenger), 187

Data about Your Activity from Partners setting, 143

Date of Birth (category), 102–103

dating, 21, 162

deactivating accounts, 126, 396

deleting
accounts, 126
from Activity Log, 151–152
albums, 278
conversations, 220, 224
friend requests, 393
fundraisers, 305–306
messages, 340
photos, 277
posts, 96

demographics of users, 31

Description field (Buy/Sell group), 299

destination icons, 37

Details About You section, 105

Developers link, 46

Device Requests icon, 163

direct messages (DMs), 288

Discussions tab (Groups), 233–234

Display and Accessibility option (Account menu), 40

Do You Want Search Engines Outside of Facebook to Link to Your Profile? setting, 128

documents, sharing, 240–241

Download Page setting (Page settings), 349

E

editing
albums, 274–275
comments, 65
descriptions, 276
Events, 315–316
fundraisers, 305–306
members of Groups, 253–255
Page info, 346
permissions, to log in as Facebook, 365
photos, 171, 267–271, 275–278, 378–379
posts, 95–96, 267–268
profile pictures, 86–87
timelines, 106

Editor (Page role), 351

education information in profile, 99–101

emails
address
in profile, 102
providing when signing up, 24
searching for friends by, 128
notifications, 25–26, 139–140
unsubscribing to, 357
valid, 23
viewing recent from Facebook, 124

Emoji option (Messenger), 219

emojis
Messenger, 189, 215–217
overview, 64
in photos, 173, 270
posting, 75

Encrypted Notification Emails, 124
End option (fundraising), 306
Engagement option (Groups), 253
Event Ticketing setting (Pages), 350
Events, 185–186
  canceling, 316
  creating, 312–315
  creating in Groups, 237–239
  deleting, 316
  duplicating, 316
  editing, 315–316
  exporting, 312
  guests, 314–317
  invitations to, 135–136, 307–310
  life events, 56
  managing, 315–317
  in News Feed, 58
  overview, 38, 307
  in Pages, 321
  posting, 56, 335
  public, 310–311
  shortcut for, 43
  viewing, 311–312
  viewing events, 185–186
Events page
  on mobile device, 162
  overview, 14
Events tab, 106

F
face books, origin of, 16
Face Recognition section (Settings page), 119, 129
Facebook. See also Events; friends; fundraisers; Games; Groups; Home page; News Feed; notifications; Pages; posting; privacy; timelines

age and, 17, 31
business promotion, 15
click-bait and, 387
community guidelines, 16–18
dating, 21
fact-checking, 387
fixing account problems, 392
frequently asked questions about, 389–396
grassroots campaigns, 386
harassment, 17–18, 388
history of, 16
interface changes, 391
job-searching, 20
logging in to mobile apps with, 366–367
logging in to websites with, 364–366
lying, 17
mobile browser, 191–194
moving, 20
other social networks vs., 18–19
overview, 7–10
paying for, 394
political activity, 383–388
Rooms feature, 19
safety, 21, 153–155
signing up for, 23–26
spam, 17–18
third-party apps, 14–15
time management, 396
Town Hall, 385–386
using, 19–21
Voting Information Center, 383–384
website for, 8, 191–192
Facebook app
  adding videos, 280–281
  avatars, 183
  commenting, 166
  Events, 185–186

Facebook Local, 191
Groups, 185–186
Instagram and, 190–191
layout of, 158–164
Messenger and, 186–191
more icon, 161–164
News Feed, 164–168
overview, 158
as part of family of mobile apps, 190–191
photos
  camera, 175–177
  editing, 269–271
  posting, 170–174
  Stories, 177–181
posting from, 168–170
profile videos, 182–183
profiles, 181–184
reacting in, 165–168
searching in, 160–161
Suggested News items, 161
WhatsApp and, 191
Facebook Community Operations team, 17
Facebook Creator Studio, 345, 362
Facebook Dating
  overview, 21
  shortcut on mobile device, 162
Facebook Language setting (Settings page), 137
Facebook Local, 191
Facebook logo (top bar), 37
Facebook Mobile, 192–194
Facebook Pages, 15
Facebook Pay. See also fundraisers
  fundraising, 301–302
  Messenger and, 189
  on mobile device, 163
  settings management, 145
  shortcut for, 44

Facebook Platform, 355–356

Facebook Portal, 220

Facebook Texts, 194–196

Facebook Watch
in News Feed, 58
overview, 21

fact-checking, 387

family information in profile, 103–104

Favorite Quotes section, 105

Featured setting (Pages), 353

Feeling/Activity button
option in share box, 169
overview, 53
posting, 333

files, sharing, 241

filters for searching, 29

Find Support option, 69

Find Wifi icon, 163

First name field (sign-up page), 24

following
friends, 208–209
posts, 242
settings management, 132

Formats for Dates, Times and Numbers setting, 137

framing profile pictures, 88–89, 377

friend requests
accepting, 31–32, 202
canceling, 201
deleting, 393
ignoring, 393
on mobile browser, 192
rejecting, 31–32
restricting, 118
sending, 201–202
settings for, 128
unwanted, 392–393

friends. *See also* users
adding
to favorites list, 167
friend requests, 201–203
to Groups, 245
overview, 31–32
choosing, 203
convincing to join Facebook, 394–395
defined, 10
following, 208–209
interacting with, 207–210
inviting, 304, 329, 361
Memories and, 381
on mobile browser, 193
muting, 167, 388
mutual, 204–205
in News Feed, 207–208
overview, 10–11, 200–201
posting in profile, 106–107
privacy settings and, 390–391
restricting, 393
reunions and, 20–21
searching for
contact information, 19
emails, 118
filtering, 29
methods for, 203–207
no results, 393–394
overview, 10, 12, 28–29
phone numbers, 118
Search box, 205–207
sharing with, 67
suggestions, 203–204
tags from, 264
unfollowing, 69
unfriending, 69, 209–210
verifying account via, 123
viewing photos from, 259–264

Friends (privacy setting), 111, 282, 308

Friends box (About section), 98

Friends Except (privacy setting), 81, 111, 283

Friends icon, 37

Friends of Friends filter, 206

Friends of Tagged checkbox (privacy setting), 113

Friends' Posts option (group), 244

Friends-only posts, 81

fundraisers, 386. *See also* Facebook Pay; Groups
creating, 302–304
deleting, 305–306
editing, 305–306
ending, 305–306
Facebook Pay and, 301–302
icon for, 78, 163
inviting friends to, 304
managing, 304–306
overview, 15–16, 38, 300–306
posts about, 59, 304–305
promoting, 304–306
shortcuts for, 43, 163
transactions and, 305
unfollowing, 305–306

G

Games
icon for, 37, 162
instant, 358–360
inviting to play, 361
in Messenger, 227
notifications for, 361
overview, 21, 358–363
posts about, 361
posts by, 59

Games *(continued)*
  removing, 141
  shortcuts for, 44, 363
  viewing activity, 363
  watching videos about, 362
  web-based, 360
Gender (category), 102–103
Gender field (sign-up page), 24
General section (Settings page), 121–122
General setting (Pages), 347–349
GIF (Graphics Interchange Format)
  commenting with, 64
  Messenger and, 189, 215
  in photos, 173, 270
  posting, 78, 169
  using in messages, 380
Group (privacy setting), 308
Group Color setting, 252
Group Quality tool, 252
Group Rules tool, 251
Groups. *See also* fundraisers; Pages
  adding descriptions to, 247–248
  adding friends to, 245
  adjusting settings, 252–253
  administering, 248–256
  Buy/Sell, 248, 297–299
  commenting in, 242–243
  controlling notifications from, 243–244
  cover photos, 247
  creating, 14, 245–248
  creating Events in, 237–239
  defined, 14
  editing members, 253–255
  evaluating, 232–235
  in Facebook app, 185–186
  interpreting insights, 253
  joining, 14, 386

  managing, 250–252
  overview, 38, 231
  pinning announcements, 249–250
  polls in, 237
  posting in, 236, 242–243, 358
  posts by, 56
  reacting to posts in, 242
  reporting offensive groups and posts, 255–256
  scheduling posts, 249
  searching, 244
  sending messages to, 213–214
  sharing and, 67, 235–237, 240–241
  shortcuts for, 44–45, 162
  types of, 248
Groups icon, 37, 159
Growth option (Groups), 253
guests
  inviting, 314–315
  messaging, 316–317
  removing, 317

H

haha icon, 62
harassment
  defined, 17
  not responding to, 388
  overview, 17–18
Help and Support
  in Account menu, 40
  on left sidebar, 46
  in Messenger, 226
  on mobile device, 164
Hidden Chats option (Messenger), 226
hiding
  comments, 130
  conversations, 224
  items from profile, 151–152

  posts and people from News Feed, 68–69, 167, 371
  sections, 106
Highlights option (Groups), 244
Home icon, 37, 159
Home option (Events), 311
Home page. *See also* News Feed
  Birthdays section, 47
  Contacts section, 47
  in Facebook Mobile, 192–194
  left sidebar, 42–46
  links on, 46
  overview, 29–31
  returning to, 37
  right sidebar, 46–47
  search bar on, 40
  shortcuts on, 44–45
  Sponsored section, 47
  top bar on, 36–40
  Your Pages section, 47
How Are You Feeling? window, 77
How People Can Find You on Facebook topic (Settings page), 118
How to Keep Your Account Secure topic (Settings page), 117

I

identity confirmation, 122
Ignore Messages option (Messenger), 219
ignoring friend requests, 393
images. *See also* posting
  3D, 377–378
  adding, 94, 173, 270, 275
  as album covers, 277
  albums
    adding to, 174
    automatic, 278
    creating, 271–273

deleting, 278
editing, 274–275
view mode, 262–263
alternative text, 267
in browser, 193
commenting and, 64
cropping, 172, 267
deleting, 277
drawing on, 173, 270
editing
  on app, 269–271
  descriptions, 276
  option for, 171
  overview, 275–278, 378–379
  while uploading, 267–268
effects, 173, 269
emojis, 173, 270
in Facebook app
  camera, 175–177
  posting, 170–174
  Stories, 177–181
from friends, 260–261
GIFs, 173, 270
icon for, 189
lighting, 173
Memories, 376
on mobile device, 263–264,
  268–269
moving, 277
in News Feed, 260–261
in Pages, 321
posting, 54, 75–76, 331–333
removing, 171
reordering, 275
rotating, 267, 276
saving, 173
scrapbooking, 376
selfies, 175
sending, 214
stickers on, 173, 270

tagging
  adding, 276
  defined, 13
  Facebook app, 172
  from friends, 264
  while uploading, 267
transferring, 125
uploading, 265–267
viewing, 259–264
inbox
  conversations in, 225–226
  in Facebook Mobile, 194
  in Marketplace, 294
  for Pages, 339–342
Incoming Call Sounds option
  (Messenger), 221
information in profile
  contact, 102–103
  Details About You section, 105
  education, 99–101
  family and relationship,
    103–104
  overview, 98
  Places Lived section, 101–102
  work, 99–101
information requests, 343
Insights section (Pages),
  343–345
Instagram
  Facebook app and, 190–191
  Facebook vs., 18
  overview, 271
Instagram setting (Pages), 353
instant emojis, 217
instant games, 141, 358–360
interacting
  with friends, 207–210
  with News Feed
    Commenting, 63–66
    common actions, 54–59

hiding posts and people,
  68–69
liking, 61–63
overview, 61
reacting, 62
saving, 67–68
sharing, 66–67
with Stories, 177–181
Interested in (category),
  102–103
Intro box (About section), 97–98
invitations
  blocking, 371
  to Events, 307–310
  to Games, 361
  to Groups, 232
  guests and, 314–315
  to Pages, 329
Issue, Electoral, or Political Ads
  setting (Pages), 352

J
jobs
  Groups for, 248
  posting as Page, 336
  searching for, 20
  shortcut for, 162
  shortcut for on sidebar, 44

K
Keyword Alerts tool
  (Groups), 252

L
languages, 102–103, 137–
  138, 349
Last name field (sign-up
  page), 24
Learn More section
  (Platform), 369

left sidebar (Home page)
 on mobile device, 161
 overview, 42–46
Legal and Policies section, 149
life events
 creating, 93–95
 overview, 56
 share box and, 90–91
Life Events box (About section), 98
Lift Black Voices icon, 163
Like button, 321
liking, 106, 344
 comments, 65–66
 overview, 61–63
 Pages, 62–63, 329, 336–337
 photos, 262
 posts, 54, 58, 242
 as reaction, 62
 unliking, 61
Limit the Audience for Posts You've Shared with Friends of Friends or Public? setting (Privacy section), 127
LinkedIn, 18
links
 in emails, 25
 on left sidebar, 46
 Messenger and, 214
 in mobile browser, 193
 in posts, 54, 75
listing (Marketplace), 290
Live Post, 334–335
Live Shopping, 289
Live Video
 Messenger and, 228
 on mobile device, 163, 169, 176
 overview, 56, 78, 281–282
location
 Messenger and, 189
 posting, 56, 78, 334
 privacy settings for, 146

Location field (Buy/Sell groups), 299
Location section (Settings page), 136
Log Out option (Account menu), 40
Log Out shortcut, 164
logging in with Facebook account
 apps, 366–367
 setting for, 123
 websites, 364–366
love icon, 62

M

mad icon, 62
Make Temporary button (profile pictures), 87
Manage Favorites option (News Feed), 208
Manage Jobs section (Pages), 342
managing
 account information, 126
 apps, 367–371
 Events, 315–317
 favorites in News Feed, 70–71
 fundraisers, 304–306
Mark as Follow Up icon, 340
Mark as Unread icon, 340
Mark as Unread option (Messenger), 223
Marketplace
 browsing, 286–289
 buying in, 286–289
 icon for, 37
 inbox in, 294
 listing in, 38
 Live Shopping, 289
 on mobile device, 161, 294–296
 posting jobs in, 292–294
 selling in, 290–292

Match Donations options (fundraising), 305
Member Requests Notifications option (Groups), 244
Member Requests tool (Groups), 251
Member-Reported Content tool (Groups), 251
Members option (Messenger), 219
Membership option (Groups), 253
Membership Questions tool (Groups), 251
memorialization settings, 122
Memories
 friends and, 381
 in News Feed, 58
 sharing, 375–376
 shortcut for, 43–44, 162
Mentorship
 icon for, 163
 parenting groups and, 248
 tool for Groups, 253
Merge Pages setting, 349
Message button, 321
message requests, 224–225
Message Requests setting (Messenger), 187, 226
Message Sounds option (Messenger), 221
messages
 blocking, 135
 deleting, 340
 guests, 316–317
 managing, 218–220
 on mobile browser, 192
 option for, 340
 replying to, 341
 using GIFs in, 380
 using stickers in, 380
 viewing, 188–189

Messaging setting (Pages), 348–350

Messenger. *See also* Rooms feature
  app version of, 228
  chat list and, 221–222
  conversations in inbox, 225–226
  Facebook app and, 186–190
  icon for, 39
  managing messages, 218–220
  message requests, 224–225
  navigating, 187–188, 222–226
  reacting in, 213
  Rooms feature, 226–228
  sending
    attachments, 217
    emojis, 215–217
    GIFs, 215
    group messages, 213–214
    links, 214
    messages, 188–189, 212–214
    payments, 216–217
    photos, 214
    stickers, 214–215
  settings for, 226
  shortcuts, 224
  video calls in, 190, 217–218
  viewing messages, 188–189

Messenger Kids
  Facebook app and, 191
  on mobile device, 163
  overview, 229

misinformation on posts, 387

mobile apps, logging in to with Facebook, 366–367

mobile browsers, Facebook on, 191–194

mobile devices
  Blood Donations section, 162
  Buy/Sell groups, 299

Crisis Response, 163
Events page, 162
Facebook Dating, 162
Facebook Pay, 163
Gaming icon, 162
Groups icon, 159
Groups page, 162
Help and Support, 164
Home icon, 159
Live Video, 163, 169, 176
Marketplace, 161, 294–296
Messenger Kids, 163
More icon, 160
Notifications icon, 159
offers, 164
Pages, 162
photos, 263–264, 268–269
Profile icon, 159
search field/icon, 160–161
Settings and Privacy, 164
shortcuts, 161–164
Town Hall, 164
Voting Information Center, 162
mobile numbers, 23–24
Mobile section (Settings page), 140
Moderator (Page role), 351
mood stories, 179
Move to Archive option (Activity Log), 151–152
Move to Archive option (Edit Post menu), 96
Move to Done icon, 340
Move to Spam icon, 340
Move to Trash option (Activity Log), 151–152
Movies icon, 163
moving photos, 277
music in Stories, 178–179
Mute Conversation option (Messenger), 219, 223

muting friends, 388
mutual friends, 204–205

N

Name Pronunciation section, 105
navigating
  Facebook app, 158–164
  Messenger, 187–188, 222–226
  profiles, 90
Nearby Friends icon, 163
New Message icon, 39
New Password field (sign-up page), 24
News Feed. *See also* Home page
  advertisements in, 59
  content in, 54–59
  controlling game posts in, 371
  defined, 11
  Events in, 58
  Facebook app, 164–168
  Facebook Watch in, 58
  friends in, 207–208
  interacting with
    commenting, 63–66
    common actions, 54–59
    hiding posts and people, 68–69
    liking, 61–63
    overview, 61
    reacting, 62
    saving, 67–68
    sharing, 66–67
  managing favorites in, 70–71
  Memories in, 58
  misinformation and, 387
  mobile browser, 192–193
  overview, 11, 52
  Pages, 323, 338–339
  parts of, 52–54

News Feed *(continued)*

  photos, 260–261

  preferences, 70–71

  refollowing and, 71

  share box, 90–91

  Snooze feature, 71

  Stories in, 59–61

  tagging in, 58

  third-party apps and, 358

  unfollowing and, 71

  viewing, 41–42

Nicknames option
  (Messenger), 219

notifications. *See also* posting

  alerts, when logging in from
    unfamiliar places, 117

  controlling from Groups,
    243–244

  email, 25–26

  Facebook texts, 196

  for Games, 361

  icon for, 159

  on mobile browser, 192

  overview, 39

  for Pages, 342–343

  push, 139–140

  of reactions, 62

  settings for, 138–140, 187,
    350–351

  third-party apps sending, 358

  toggling, 96

O

Off option (Groups), 244

Off-Facebook Activity, 125–126

Off-Facebook Previews
  setting, 132

one-time passwords, 195

Only Me (privacy setting), 81,
  111, 283

Open Chat option
  (Messenger), 228

Open in Messenger option, 218

Options icon (Messenger), 39

Other Names section, 105

Other Tagging This Page setting
  (Pages), 348

P

Page Engagement (Insight
  section), 344

Page Info setting (Pages), 350

Page Likes (Insight section), 344

Page Management History
  setting (Pages), 354

Page Moderation setting
  (Pages), 348

Page Quality section
  (Pages), 346

Page Roles setting (Pages),
  351–352

Page Support Inbox setting
  (Pages), 354

Page Update setting, 349

Page Visibility setting, 347

Pages. *See also* Groups

  Ad Center section, 346

  automated responses,
    341–342

  blocking, 136

  Business App Store
    section, 342

  call-to-action button, 329

  commenting, 336–337

  connecting with, 322–324

  cover photos, 320–321, 327

  creating, 325–329

  Edit Page Info section, 346

  inbox of, 339–342

  Insights section, 343–345

  interacting with, 322–324

  inviting friends to, 329

  liking, 329, 336–337

  Manage Jobs section, 342

  message options for, 340

  More icon on, 321

  News Feed of, 338–339

  notifications for, 342–343

  overview, 15, 38, 320–324

  Page Quality section, 346

  parts of, 320–322

  posting

    advertisements, 336

    Events, 335

    feeling/activity info, 333

    jobs, 336

    Live Post, 334–335

    location, 334

    offers, 335–336

    photos, 331–333

    promoting, 334

    scheduling, 334

    share box, 329–334

    specialty posts, 334–336

    videos, 331–333

  profile pictures, 320–321,
    326–327

  Publishing Tools section,
    345–346

  purpose of, 324–325

  replying to messages, 341

  Resources & Tools section, 342

  search icon, 321

  setting up, 328

  settings management

    Activity Log setting, 354

    Ad Limits setting, 352

    Advanced Messaging
      setting, 351

    Branded Content setting,
      352–353

    Crossposting setting, 354

    Event Ticketing setting, 350

    Featured setting, 353

    General setting, 347–349

    Instagram setting, 353

Issue, Electoral, or Political
  Ads setting, 352
Messaging setting, 349–350
Notifications setting, 350–351
Page Info setting, 350
Page Management History
  setting, 354
Page Roles setting, 351–352
People and Other Pages
  setting, 352
Preferred Page Audience
  setting, 352
Templates and Tabs
  setting, 350
WhatsApp setting, 353
sharing to, 67
shortcuts for, 43, 162
tabs, 321
URLs, 328
viewing, 329
Part option (Account menu), 39
passwords
  changing, 122–123
  Facebook texts and, 195
  option to remember, 40
  overview, 117
  providing upon sign-up, 24
paying for Facebook, 394
payments, sending, 216–217
Pending Posts tool (Groups), 251
People and Other Pages setting
  (Pages), 352
People icon (Messenger), 188
People You May Know
  section, 10
personal information, apps
  accessing, 356–358, 369
phishing, 154–155, 220
phone numbers, 102, 128
Photo field (Buy/Sell group), 299
photo viewer, 165, 261–262

photos. See also posting
  3D, 377–378
  adding, 94, 173, 270, 275
  as album covers, 277
  albums
    adding to, 174
    automatic, 278
    creating, 271–273
    deleting, 278
    editing, 274–275
    view mode, 262–263
  alternative text, 267
  in browser, 193
  commenting and, 64
  cropping, 172, 267
  deleting, 277
  drawing on, 173, 270
  editing
    on app, 269–271
    descriptions, 276
    option for, 171
    overview, 275–278, 378–379
    while uploading, 267–268
  effects, 173, 269
  emojis, 173, 270
  in Facebook app
    camera, 175–177
    posting, 170–174
    Stories, 177–181
  from friends, 260–261
  GIFs, 173, 270
  icon for, 189
  lighting, 173
  Memories, 376
  on mobile device, 263–264,
    268–269
  moving, 277
  in News Feed, 260–261
  in Pages, 321
  posting, 54, 75–76, 331–333

removing, 171
reordering, 275
rotating, 267, 276
saving, 173
scrapbooking, 376
selfies, 175
sending, 214
stickers on, 173, 270
tagging
  adding, 276
  defined, 13
  Facebook app, 172
  from friends, 264
  while uploading, 267
transferring, 125
uploading, 265–267
viewing, 259–264
Photos box (About section), 98
Photo/Video option (Facebook
  app), 169
Pinterest, 364
Places Lived section, 101–102
PMs (private messages), 288
Political Views (category),
  102–103
polls, 179, 237, 310, 332
Pop-up New Messages option
  (Messenger), 221
Portal, 220
Post and Story Sharing setting
  (Pages), 348
Post in Multiple Languages
  setting (Pages), 349
Post Reach (Insight section), 344
Post Topics (Groups), 251
posting. See also notifications;
  photos; share box; videos
  about comments, 58
  about cover photos, 58
  about fundraisers, 304–305
  about fundraising, 59

posting (continued)
  about Games, 361
  about liking, 58
  about location, 56
  about profile pictures, 58
  activities, 77
  Add to Your Post window
      and, 78
  by apps, 59
  archiving posts, 96
  backgrounds, 74–75
  commenting in posts, 54
  emojis, 75
  Events, 56
  from Facebook app
      options, 166–168
      overview, 168–170
      photos, 170–174
  following posts, 242
  by Games, 59
  GIFs, 78
  in Groups, 56, 236, 242–
      243, 358
  hiding posts, 167, 371
  How Are You Feeling?
      window, 77
  jobs in Marketplace, 292–294
  liking posts, 54, 242
  links, 54, 75
  location, 78
  Memories, 376
  in News Feed, 358
  overview, 12–13, 38, 71–72
  in Pages
      advertisements, 336
      Events, 335
      feeling/activity info, 333
      jobs, 336
      Live Posts, 334–335
      location, 334
      offer, 335–336
      photos, 331–333
  promoting, 334
  scheduling, 334
  share box, 329–334
  specialty posts, 334–336
  photos, 54, 75–76
  pinning posts, 249–250
  previews of websites in
      posts, 75
  privacy settings, 80–82
  in profile
      editing, 95–96
      friends, 106–107
      overview, 90–93
      third-party apps, 358
  profile picture in posts, 53
  reacting to posts, 62
  reporting posts, 167, 255–
      256, 387
  restrictions on, 18
  saving posts, 166
  scheduling, 249
  sharing posts, 54
  specialty posts, 334–336
  sponsored posts, 59
  status updates, 72–73
  Stories, 78–80
  suggested posts, 59
  tagging, 54, 76–77, 380
  translating posts, 137–138
  unfollowing posts, 242–243
  videos, 54, 75–76
  from websites, 366
Posts and Stories (Settings
    page), 116
Preferred Page Audience setting
    (Pages), 352
Preview mode (Groups), 232
Price field (Buy/Sell group), 299
privacy, 46, 54
  for advertisements, 142–145
  for albums, 274–275
  Apps and Websites
      section, 140
  audience, 110–113
  Blocking section, 133–136
  Business Integrations
      section, 142
  Face Recognition section, 129
  Facebook Pay section, 145
  friends and, 390–391
  General section, 121–122
  How People Can Find You on
      Facebook topic, 118
  How to Keep Your Account
      Secure topic, 117
  Instant Games section, 141
  Language and Region section,
      137–138
  Location section, 136
  Mobile section, 140
  Notifications section, 138–140
  overview, 29, 282–284
  for photos, 282–284
  for posting, 80–82
  Privacy section, 127–128
  for profile, 114–115
  Profile and Tagging section,
      129–131
  Public Posts section, 131–133
  restricted list and, 133–134
  Security and Login section,
      122–124
  under share box, 113–114
  for Stories, 138
  Support Inbox section, 145
  for videos, 146, 282–284
  Who Can See What You Share
      topic, 116–117
  Your Ad Preferences on
      Facebook topic, 119–120
  Your Data Settings on
      Facebook topic, 119
  Your Facebook Information
      section, 124–126
Privacy Checkup page, 115–120
Privacy section (Settings page),
    127–128

Privacy Shortcuts page
  Activity Log, 150–152
  overview, 146–149
  reporting and, 152–153
  View As tool, 150
Private (privacy setting), 308
private and visible/hidden
  groups, 233
private messages (PMs), 288
Product Tags field (Buy/Sell
  groups), 299
Profanity Filter setting
  (Pages), 348
Profile and Tagging section
  (Settings page), 129–131
Profile icon, 159
Profile Information section
  (Settings page)
  overview, 116
  personalized advertisements
    and, 119–120
profile pictures
  adding, 26–28, 326–327
  defined, 84
  editing, 86–87
  framing, 88–89, 377
  on mobile browser, 193
  for Pages, 320–321
  in posts, 53
  posts about, 58
  on top bar, 38
profile videos, 182–183
profiles. See also accounts
  adding information to, 32–33
  cover photos, 85–86
  creating posts in, 90–93
  defined, 11
  editing posts in, 95–96
  elements on left column of,
    96–98
  emails on, 102
  in Facebook app, 181–184

in Facebook Mobile, 193
first impressions and,
  84–89
friends and, 106–107
information in
  bio, 89
  contact, 102–103
  Details About You
    section, 105
  education, 99–101
  family and relationship,
    103–104
  overview, 98
  Places Lived section,
    101–102
  work, 99–101
life events in, 93–95, 105
miniature version of, 53
navigating, 90
overview, 11–12
Pages, 322
posting in, 358
privacy settings, 114–115
rumor about, 390
share box, 90–91
tabs, 105–106
viewing, 390
websites in, 102
promoting
  businesses, 15
  fundraisers, 304–306
  posts, 334
protecting minors, 154
Public (privacy setting), 80,
  110–111, 282, 308
public and visible groups, 232
public events, 310–311
Public Post Comments/
  Notifications settings
  (Public Posts section), 132
Public Posts section (Settings
  page), 131–133

Publishing Tools section (Pages),
  345–346
push notifications, 139–140

R
reacting to posts
  comments
    adding, 63–65
    censoring words from, 130
    editing, 65
    GIFs and, 64
    in Groups, 242
    liking, 65–66
    Pages and, 336–337
    photos and, 64
    in posts, 54, 166
    posts about, 58
    removing, 65
    replying to, 65–66
    stickers, 64
    tagging, 64–65
  defined, 62
  in Facebook app, 165–168
  in Groups, 242
  in Messenger, 213
  overview, 62, 165–168
reading posts, 242–243
Recent Ad Activity icon, 164
reciprocal friendships, 31
recommendation requests, 58
Reconnect option (News
  Feed), 208
recovering accounts, 124
re-following, 71
rejecting friend requests,
  31–32
Related Pages box, 323
relationship information in
  profile, 103–104
Religion (category), 102–103
Remember Password option, 40

Remove from Page Likes setting, 352
Remove Page setting, 349
removing
  apps, 369–370
  games, 141
  guests, 317
  photos, 171
  tags, 264
re-ordering photos, 275
replying
  to comments, 65–66
  to messages, 341
reporting
  apps, 371
  Groups, 255–256
  overview, 152–153
  posts, 69, 255–256, 387
Resources & Tools section (Pages), 342
restricting
  friends, 393
  restricted list, 133–134
  tags, 130
Review All Your Posts and Things You're Tagged In setting (Privacy section), 127
Review Posts Friends Tag You in before They Appear on Your Profile? setting (Profile and Tagging section), 284
Review Posts You're Tagged in Before the Post Appears on Your Profile setting (Profile and Tagging section), 131
right sidebar, 46–47
Rooms feature. *See also* Messenger
  Groups and, 252
  Messenger and, 189, 226–228
  overview, 19, 38
rotating photos, 267, 276

S

sad icon, 62
Safety Check feature, 21, 44
safety on Facebook, 153–155, 296
Safety section, 149
Save feature, 67–68
Save Post option (Edit Post menu), 95
Save Your Login Info section, 123
Saved (shortcut in sidebar), 44, 162
saved replies, 341
Scheduled Posts (Groups), 251
scheduling posts, 249, 334
scrapbooking photos, 376
search engines, 118
search field/icon
  Messenger, 39
  on mobile browser, 193
  on mobile device, 160–161
  overview, 40
  in Pages, 321
  top bar, 37
Search page, 206–207
securing account, 148
Security and Login section (Settings page), 122–124
See All in Messenger icon, 39
selfies, 175, 179
selling
  in Buy/Sell group, 298–299
  items in Facebook app, 170
  in Marketplace, 290–292
Send in Messenger option, 67
sending friend requests, 201–202
Settings and Privacy
  option in Account menu, 39
  shortcut on mobile device, 164

settings for Messenger, 226
share box. *See also* posting
  creating poll with, 237
  in Facebook app, 168–170
  on mobile browser, 193
  overview, 72
  Pages, 322, 329–334
  posting with, 235–237
  privacy settings under, 113–114
  in profile vs. in News Feed, 90–91
Share Now option, 66
sharing
  content, 366
  documents, 240–241
  feature for, 66–67
  files, 241
  with friends, 67
  with Groups, 67, 235–237, 240–241
  Memories, 375–376
  Pages, 67
  photos, 13–14
  posts, 54
  videos, 13–14, 175–177
Shop icon, 162
shortcuts on Home page, 44–45
Show Contact option (Messenger), 222
signing up for Facebook, 23–26
Similar Page Suggestions setting (Pages), 349
SMS notifications, 139–140
Snapchat, 18
Snooze feature, 69, 71, 167, 208
Social Interactions setting (Ad Settings section), 145
Social Interactions setting (Settings page), 120
social learning groups, 248

social networks, 18–19

Something's Wrong option (Messenger), 220

spam, 17–18, 154

specialty posts, 334–336

Specific Friends (privacy setting), 81, 112, 283

sponsored posts, 59

Sponsored section, 47

status updates
  backgrounds, 379–380
  content of, 73
  defined, 53
  Memories, 376
  overview, 54, 72–73
  voting, 384–385

stickers
  commenting and, 64
  Messenger and, 189, 214–215
  in photos, 173, 270
  using in messages, 380

Stories
  Boomerang option for, 179
  in Facebook app, 177–181
  interacting with, 177–181
  mood option, 179
  music in, 178–179
  in News Feed, 59–61
  overview, 38
  polls on, 179
  posting, 78–80
  privacy settings for, 138
  selfies and, 179
  text in, 178
  viewing, 41, 177–181

Story Archive tab, 106

subscribing to Pages, 323

Suggested News item (Facebook app), 161

suggested posts, 59

Superzoom option, 176

Support Inbox section (Settings page), 145

Support Nonprofit option (Facebook app), 170

Switch Account option (Account menu), 39

T

Tag Friends option (Facebook app), 169

tagging. See also notifications
  branded content and, 353
  commenting and, 64–65
  defined, 13
  in News Feed, 58
  overview, 76–77
  photos, 172, 262, 264, 276
  posts, 54, 380
  restrictions on, 130

Tagging Ability setting (Pages), 348

Templates and Tabs setting (Pages), 350

Text Messaging setting (Facebook Text), 196

third-party apps
  blocking, 371
  hiding posts about, 371
  overview, 14–15
  posting on Facebook, 358
  sending notifications, 358

thumbnails for videos, 332

time management, 396

Time of Day setting (Facebook Text), 196

timelines. See also accounts
  adding information to, 32–33
  cover photos, 85–86
  creating posts in, 90–93
  defined, 11
  editing posts in, 95–96

elements on left column of, 96–98

emails in, 102

in Facebook app, 181–184

in Facebook Mobile, 193

first impressions and, 84–89

friends and, 106–107

information in
  bio, 89
  contact, 102–103
  Details About You section, 105
  education, 99–101
  family and relationship, 103–104
  overview, 98
  Places Lived section, 101–102
  work, 99–101

life events in
  creating, 93–95
  section with, 105

miniature version of, 53

navigating, 90

overview, 11–12

Pages, 322

privacy settings for, 114–115

rumor about, 390

share box in, 90–91

tabs, 105–106

third-party apps and, 358

viewing, 390

websites in, 102

timestamp, 54

Title field (Buy/Sell group), 299

top bar (Home page), 36–40

Town Hall, 44, 164, 385–386

tracking labels, 332

transactions, fundraisers and, 305

transferring media from account, 125

Translate Automatically setting (Pages), 349

translating posts, 137–138

trolling. *See* harassment

Turn Off Active Status option (Messenger), 222

Turn Off/On Notifications for This Post option (Edit Post menu), 96

Turn Off/On Translations option (Edit Post menu), 96

Turn On Notifications for This Post option, 168

Twitter, 18

Two-Factor Authentication, 117, 123

U

Unfollow option (News Feed), 208

unfollowing
  blocking and, 136
  in Facebook app, 167
  fundraisers, 305–306
  Groups, 244
  News Feed and, 71
  overview, 69
  posts, 242–243

unfriending, 134, 209–210

Unlike option (Activity Log), 151–152

unliking, 61, 136

uploading
  editing photos while, 267–268
  photos, 265–267
  photos on mobile device, 268–269
  videos, 280–281

URL for Pages, 328

usernames, 121

users. *See also* friends
  banning from Pages, 352
  blocking, 116, 134–135

demographics of, 31

in Groups, editing, 253–255

restricting, 133–134

unfriending vs. unfollowing, 69

V

valid emails, 23

valid mobile numbers, 23

verified Pages, 321

video calls, 190, 217–218

video chat, 38

Video Chat option (Messenger), 224

videos. *See also* photos; posting
  3D, 377–378
  adding, 94, 280–281
  auto-play, 146
  boomerang, 176
  captions, 146, 332
  as cover photos, 327
  enabling sound, 165
  Live Video, 281–282
  in Messenger, 228
  polls on, 332
  in posts, 54
  privacy settings for, 146
  as profile picture, 182–183
  settings management, 146
  share box, 169
  sharing, 13–14, 175–177
  thumbnails, 332
  transferring, 125
  uploading, 280–281
  viewing, 279
  Year in Review, 379

Videos tab, 106

View As tool, 150

View Profile option (Messenger), 218, 223

viewing
  Events, 311–312

gaming activity, 363

News Feed, 41–42

Pages, 329

photos, 259–264

profiles, 390

Stories, 41

videos, 279, 362

Viewing and Sharing section, 130

viruses, 154

Visitor Posts setting (Pages), 348

voice messages, 189

voting, status updates about, 384–385

Voting Information Center
  on mobile device, 162
  overview, 383–384

W

Watch feature, 37–38, 59, 78, 169

Weather icon, 164

web address setting (Group), 252

web cookies, 46

web-based games, 360

websites
  Facebook
    computer, 8
    for mobile browser, 191–192
  logging in to with Facebook account, 364–366
  previews of, 75
  in profile, 102

WhatsApp, 191, 353

When You're Tagged in a Post, Who Do You Want to Add to the Audience If They Can't Already See It? setting (Settings page), 130–131, 284

Where You're Logged In section, 122

Who Can Follow Me setting (Settings page), 132

Who Can Join the Group setting (Settings page), 252

Who Can Post on Your Profile? setting (Settings page), 130

Who Can See Posts You're Tagged in on Your Profile? setting (Settings page), 130

Who Can See Posts You've Been Tagged in on Your Profile? setting (Settings page), 283

Who Can See the People, Pages and Lists You Follow? setting (Settings page), 127

Who Can See What Others Post on Your Profile? setting (Settings page), 130

Who Can See What You Share topic (Settings page), 116–117

Who Can See Your Future Posts? setting (Settings page), 127

Who Can Send You Friend Requests? setting (Settings page), 128

Why Am I Seeing This Post? option (Settings page), 167–168

work groups, 248

wow icon, 62

Write Post option, 66

Y

Year in Review videos, 379

Your Ad Preferences on Facebook topic (Settings page), 119–120

Your Data Settings on Facebook topic (Settings page), 119

Your Events option, 311

Your Facebook Information section (Settings page), 124–126, 148–149

Your Messenger URL option (Pages), 349

Your Pages section, 47

Z

Zuckerberg, Mark, 16

About the Authors

Carolyn Abram is a writer. She was the first user of Facebook at Stanford in 2004 and worked for Facebook from 2006–2009. She has used Facebook every day for most of her adult life, acquiring 805 Facebook friends. Despite that, she managed to receive a BA from Stanford ('06) and an MFA from California College of the Arts ('12). Her short fiction has appeared in *New California Writing 2013*, *Switchback*, and *The Offbeat*. Her essays have appeared in *McSweeney's Internet Tendency* and *Lilith* online. She currently lives in Seattle with her husband, children, and dog.

Amy Karasavas is a sommelier. She was an early employee of Facebook, working there from 2007 to 2010. She currently leads wine tastings and classes through her company, That Pour Woman. She lives with her partner Michael and their two dogs in Brooklyn, NY.

Authors' Acknowledgments

Carolyn: A huge thank you to Amy Karasavas for signing on to working on this edition; your feedback and comments have been invaluable. Thanks to the flexibility and support of the team of editors — especially Susan and Michelle — who have been so wonderful to work with in a not very wonderful year.

On the home front, I want to thank Eric, Connor, and Lina for being themselves. While at times the quarantine walls have felt as though they were closing in around us, I am very, very glad to have you all in here with me. Although it was very nice when you left my office so I could get some work done.

In closing, I'd like to thank the billions of Facebook users around the world who are busy connecting, sharing, and generally having fun on Facebook. Keep on signin' on.

Amy: I can't say thank-you enough to editors Susan Pink and Michelle Krazniak for their assistance and their wisdom throughout this process. We make quite the team! Also a huge thank-you to the Wiley crew who have guided this edition as well as past editions. Carolyn and I have known each other since our days at Facebook and she continues to be a wonderful friend and co-author.

Endless gratitude to my partner, Michael, for his love, his jokes, and his continual support. And to our two adopted dogs, Baco and Porthos — they fill our lives with mayhem and delight.

Lastly, I'm thankful to have been a part of Facebook's early days and it's my sincere hope that the company continues to evolve and address the ways in which it has fallen short for its users. At its best, Facebook is all about connection, empathy, and having a positive effect.

Publisher's Acknowledgments

Executive Editor: Steve Hayes
Project Editor: Susan Pink
Copy Editor: Susan Pink
Technical Editor: Michelle Krazniak
Proofreader: Debbye Butler

Production Editor: Tamilmani Varadharaj
Cover Image: © Urupong/iStock/Getty Images